Researching Peacebuilding in Africa

This book examines the multifaceted nature of conflict and the importance of the socio-economic and political contexts of conflict and violence and shows how to support ongoing initiatives and programs to build sustainable peace on the African continent.

Drawing on a range of conceptual framings in the study of peace and conflict, from gender perspectives to institutionalist to decolonial perspectives, the contributors show how peacebuilding research covers a whole range of questions that go beyond concerns for post-conflict reconstruction strategies. Chapters focus on the methodological, theoretical and practical aspects of peacebuilding and provide a toolbox of perspectives for conceptualizing and doing peacebuilding research in Africa. Anchored in African-centered perspectives, the book encourages and promotes high-quality interdisciplinary research that is conflict-sensitive, historically informed, theoretically grounded and analytically sound.

This book will be of benefit to scholars, policy makers and research institutions engaged in peacebuilding in Africa.

Ismail Rashid is a Professor of History and International Studies at Vassar College, USA.

Amy Niang is a Senior Lecturer in International Relations at the University of the Witwatersrand, South Africa.

Routledge Studies in Peace, Conflict and Security in Africa
Edited by Cyril Obi, Social Science Research Council, New York, USA

1 **The International Criminal Court and Peace Processes in Africa**
Judicialising Peace
Line Engbo Gissel

2 **The Unfinished Revolution in Nigeria's Niger Delta**
Prospects for Socio-economic and Environmental Justice and Peace
Edited by Cyril Obi and Temitope B. Oriola

3 **Developmental Regionalism and Economic Transformation in Southern Africa**
Edited by Said Adejumobi and Cyril Obi

4 **Researching Peacebuilding in Africa**
Reflections on Theory, Fieldwork and Context
Edited by Ismail Rashid and Amy Niang

Researching Peacebuilding in Africa

Reflections on Theory, Fieldwork and Context

Edited by
Ismail Rashid and Amy Niang

Routledge
Taylor & Francis Group
LONDON AND NEW YORK

First published 2021
by Routledge
2 Park Square, Milton Park, Abingdon, Oxon OX14 4RN

and by Routledge
52 Vanderbilt Avenue, New York, NY 10017

Routledge is an imprint of the Taylor & Francis Group, an informa business

British Library Cataloguing-in-Publication Data
A catalogue record for this book is available from the British Library

Library of Congress Cataloging-in-Publication Data
A catalog record has been requested for this book

ISBN: 978-0-367-90411-1 (hbk)
ISBN: 978-1-003-11174-0 (ebk)

Typeset in Bembo
by codeMantra

Contents

Illustrations

Figures

Tables

Contributors

Dr. Festus Kofi Aubyn is a Research Fellow at the Kofi Annan International Peacekeeping Training Centre (KAIPTC) in Ghana. He is also an Adjunct Lecturer at the Webster University and the Ghana Armed Forces Staff and Command College. He served as a visiting Lecturer at the Department of Political and International Studies, Rhodes University, South Africa in 2016. His research focuses on African peace and security with a particular focus on peace operations, terrorism and election security. He has published widely and presented papers in a number of international conferences on these subjects.

Asebe Regassa Debelo is an Associate Professor at Dilla University, Ethiopia and was a postdoctoral fellow at the University of Zürich, Switzerland (2015–2016). He has served as director of Research and Dissemination Office, Dilla University from 2017 to 2019. Over the last couple of years, Dr Debelo has been doing research on conflict, violence, peace-building and community resilience. He extensively published peer-reviewed journal articles and book chapters, policy brief notes and commentaries on diverse ranges of issues including inter-ethnic conflict, state consolidation, nature conservation, indigenous peoples' rights, indigenous notions of peace, displacement and the political economy of mega-development projects in Ethiopia. His research areas mainly cover the pastoralist frontiers in the Horn of Africa.

Olayiwola Erinosho, a sociologist by training, received his education at the Universities of Ibadan, Nigeria and Toronto, Canada. Professor Erinosho previously served as Executive Secretary, Social Science Academy of Nigeria and of the Health Reforms Foundation of Nigeria. He is Fellow of the Social Science Academy of Nigeria.

Heidi Hudson is a Professor of International Relations and Dean of the Faculty of the Humanities at the University of the Free State, Bloemfontein, South Africa. She specializes in feminist security studies, with a specific focus on Africa. Her research interests concentrate on discursive and material dynamics of peacebuilding in the postcolony. From 2011 to

2017, she was co-editor of the *International Feminist Journal of Politics* – a journal that operates at the intersection of politics, international relations and women's studies. She is currently an Advisory Board member of the African Peacebuilding Network of the Social Science Research Council in New York. In 2018, she was the Claude Ake Visiting Chair at the Nordic Africa Institute and the Department of Peace and Conflict Research in Uppsala, Sweden. Heidi Hudson has published in journals such as *International Peacekeeping, Peacebuilding, Security Dialogue, Security Studies* and *Politics & Gender.*

Dr. Fatma Osman Ibnouf is currently an Assistant Professor at Development Studies and Research Institute, University of Khartoum, Sudan. She holds a PhD degree from University of Wales Swansea, UK. She is the author of several articles, book chapters, and book. Her most recent book is: *War-Time Care Work and Peacebuilding in Africa: The Forgotten One* (Palgrave Macmillan, 2020). Her research interests lie in gender/women, food Security, care-work and peacebuilding in Sudan, Africa, and Arab World. Her awards include Student Grant from Arab British Chamber of Commerce, London, UK; Special Research Grant from Finnish Ministry of Foreign Affairs, Institute for Human Rights at Åbo Akademi University, Finland, and other fellowships. She is a 2017 recipient of the Social Science Research Council's African Peacebuilding Network's Book Manuscript Completion Grant and a 2016 recipient of its Individual Research Grant. Dr. Fatma is a member of Collaborative Working Group Research Fellowship of the APN, SSRC-2018–2020.

Dr. Jimam T. Lar is currently a Lecturer in History at the University of Jos. He received his BA and MA at University of Jos, MA in Conflict, Security and Development from King's College London, and a PhD in African History and Politics from the University of Bayreuth, Germany. Dr Lar has done extensive research on the history of security provisioning (plural policing) and violence in Nigeria, and conflict management and peacebuilding in West Africa. He was a recipient of the African Peacebuilding Network (APN), Social Science Research Council of New York, Individual Grant Recipient (2016) and an African Humanities Program (AHP) Postdoctoral Research Fellow (2017–2018). Dr Lar has published several journal articles and book chapters. His current research is focused on the post conflict transformation of community based armed groups, the role of local actors and trans-border security, mobility on the Nigerian-Nigerien and the Nigeria-Cameroon borderlands, and New Methodologies in African History: Theories, Concepts and Multidisciplinarity.

Pamela Chepngetich Mainye is currently a Lecturer and the Head of Department, Communication and Journalism Studies, at the faculty of Information Science & Technology at Kisii University, Kenya. Dr Mainye holds a PhD in Media Studies from the University of Bayreuth, Germany, where she was a junior fellow at the Bayreuth International Graduate

School of African Studies (BIGSAS). Her doctoral thesis there focused on self-representation of refugees in Kenya's Daadab Refugee Camp. Her other research interests include livelihoods, social media and online activism as well as radio studies. She was a 2017 APN grantee where she did research on how radio fan clubs are convened, and organically form by bringing people of different ethnic, class and political persuasions together, and which ultimately plays the role in sustaining cohesion and everyday human security in Western Kenya.

Godfrey Maringira is an Associate Professor of Anthropology at Sol Plaatje University, South Africa. He is also a Research Associate at the University of Johannesburg. He is a senior Volkswagen Stiftung Foundation research fellow. He is also a Principal Investigator of the International Development Research Center, Canada (IDRC) research project and a two-time consecutive recipient of the SSRC's Next Generation Social Sciences in Africa and a three-time recipient of the African Peacebuilding Network grants. He is currently serving as the Board member of the Next-Generation Social Sciences in Africa. His areas of research include the military and armed violence in Africa. He was awarded the 2018 *African Affairs* Journal best author price. In 2020, he was awarded another author best price: The Benedict Wallet Bambatha Vilakazi Prize in the *African Studies* Journal. His book *Soldiers and the State in Zimbabwe* was published in 2019 by Routledge.

Amy Niang teaches International Relations at the University of the Witwatersrand. She is a Visiting Professor at the University of Sao Paulo and the Institute for Peace and Security Studies (IPSS) in Addis Ababa. She is the author of *The Postcolonial African State in Transition: Stateness and Modes of Sovereignty* (Rowman and Littlefield 2018).

Duncan Omanga is currently Program Officer of the Africa Peace Network (APN) and Next Generation Social Sciences, at the Social Science Research Council, Brooklyn. Dr Omanga is also a senior lecturer of Media Studies at Moi University (on leave) and a former British Academy visiting fellow at the University of Cambridge (2018/2019). He is an alumnus of the Bayreuth International Graduate School of Graduate Studies (BIGSAS) at the University of Bayreuth, Germany. Dr Omanga is also a 2014 APN (Africa Peace Network-SSRC) alumni and was the 2015/2016 African Visiting Fellow (Centre for African Studies) at the University of Cambridge. Dr Omanga has researched and published on East Africa's digital and print media. His current research interests are on security, digitality and policing practices in Kenya.

Kenneth Omeje is Visiting Professor at the Institute for Peace and Security Studies in Addis Ababa University, Ethiopia; Visiting Professorial Fellow at the Nigerian Defence Academy, Kaduna, Nigeria; and Director, Manifold Crown Consulting Services – an international peacebuilding and security policy intervention consult based in Bradford, UK. He has previously held the positions of Professor of International Relations at the United States International University in Nairobi, Kenya; Senior Research Associate, Faculty of Humanities, University of Johannesburg, South Africa; and Senior Research Fellow at the John and Elnora Ferguson Centre for African Studies, University of Bradford, UK. Kenneth has more than 90 publications, including books, book chapters, policy research papers, contributions to international encyclopedias and articles in well-regarded journals. His most recent book is titled *Peacebuilding in Contemporary Africa: In Search of Alternative Strategies* (edited, London: Routledge, 2019).

Ismail Rashid is a Professor of History and International studies at Vassar College since 1998. His primary teaching interests are pre-colonial and modern African history, African Diaspora and Pan-Africanism, and International Relations. His research interests include subaltern resistance against colonialism, public health, and conflicts and security in contemporary Africa. Among his recent books are *West Africa's Security Challenges* (2004 with Adekeye Adebajo), *The Paradox of History and Memory in Postcolonial Sierra Leone* (2013) (with Sylvia Ojukutu-Macauley) and *Understanding West Africa's Ebola Epidemic: Towards a Political Economy*(2017) (with Ibrahim Abdullah).

Dr. Norman Sempijja is a Senior Lecturer at the University of Navarra in Pamplona, Spain. He holds a PhD. in international politics from Kingston University, a Masters in Diplomatic Studies from Keele University and BA Social Sciences from Makerere University. His research interests are in security studies where he focuses on inter-organizational cooperation, peacekeeping, peace enforcement and peacebuilding, securitization and security-development nexus. Key geographical areas of research focus are but not limited to mainly Africa and Europe. His teaching expertise covers World politics, European Union Security, African international relations, international security and international organizations among others.

Thomas Kwasi Tieku is an Associate Professor of Political Science at King's University College at The University of Western Ontario in Canada. He is the former Director of African Studies at the University of Toronto where he won the Excellence of Teaching Award. He has coordinated the Social Justice and Peace Studies program at King's, served as the Lead Researcher at the Centre for International Governance Innovations (CIGI) and was 2017 Carnegie Fellow at the University of Ghana, Legon. Thomas' current research, which is supported by Canada's Social Sciences and Humanities Research Council (SSHRC), focuses on informal international relations, mediation, peacebuilding and international organizations. He has authored, co-authored or co-edited four (4) books and over thirty (30) refereed journal articles and book chapters. He has served as a consultant to several organizations and governments, including the World Bank Group, UN, AU, CIGI and the Government of Canada.

Foreword

As we enter the age of second-generation peacebuilding knowledge production, *Researching Peacebuilding in Africa: Reflections on Theory, Fieldwork and Context* brings together critical reflections on the nexus between theory, method and practice. These reflections are mainly from former grantees and mentors involved in different methods, writing and dissemination workshops organized by the Social Science Research Council's African Peacebuilding Network (APN) program, with support from the Carnegie Corporation of New York (CCNY). Launched in 2012, the APN has been providing funding and research capacity building support for African scholars to engage in field-based research aimed towards producing high quality and globally visible knowledge on African peacebuilding. In this regard, the program has successfully transformed the thinking, careers and lives of its grantees. Since the program's inception, they have progressively transformed the discourse of African peacebuilding through a series of innovative research projects and publications exploring the nexus between conflict, peace and security across all of Africa's regions.

Co-edited by the former Chair of the APN's advisory board, Ismail Rashid, and Amy Niang, a current member of the APN's advisory board and a former APN grantee, this volume features the work of APN grantees and mentors, and speaks eloquently to the dynamic intergenerational dialogues on African peacebuilding facilitated over the years by the program. It is emblematic of the positive impact of the APN—its grantees, mentors and burgeoning network, representing a distinct voice on innovative knowledge production on African peacebuilding. The volume is a significant representation of the excellent contributions of APN grantees—whose works continue to expand the scope and scale, as well as the quality and visibility of rapidly expanding African scholarship in the field of peacebuilding.

Such contributions have also played a considerable part in drawing attention to hitherto marginalized or neglected aspects of peacebuilding in Africa, ranging from the role of new or emerging actors in conflict, and emerging issues related to wartime care work, digital and social media, gender, post-conflict memories, trauma and reconciliation, the role of diaspora communities, global-regional-local connections, as well as the transboundary dimensions of conflict and peace in Africa, among others. Some of their works have also opened up new vistas to the connections between natural resources, conflict and peace, local-national-regional peace nexus and the linkages between communities-led initiatives, citizenship, statebuilding and peacebuilding. Equally noteworthy is how some grantees have deployed their research-based knowledge and activism towards shaping contents of curricula, policy and practice in the field of African peacebuilding.

The collection of essays in this volume represents a refreshing addition to the rapidly growing field and scholarship on African peacebuilding that is capturing and unpacking the shifts in the landscape of conflict and security, peace and development on the continent. Equally important is how the chapters cumulatively point the way forward by deploying an interdisciplinary and reflexive approach towards a truly African perspective to peacebuilding. Overall, the book fundamentally demonstrates the significance of the APN network and its potential to redefine and shape more productive relations between research and policy in African peacebuilding, in ways that also capture the important work and contributions of members of successive cohorts of the program's grantees. It will stimulate new thinking and a paradigm shift in research, policy and practice in ways that reassert Africa's contribution to global discourses on peacebuilding.

Cyril Obi
Program Director, African Peacebuilding Network and
Next Generation Social Science in African program,
Social Science Research Council (SSRC), New York.

Introduction

Ismail Rashid and Amy Niang

Despite the exuberance and emancipatory aspirations of independence from European colonialism that began in earnest six decades ago, African governments and societies have been wrestling with the challenge of forging stable political, social, and economic systems that are responsive to the interests, needs, and aspirations of their diverse populations. The various structures of postcolonial governance, whether indigenous, inherited from colonialism, or created after independence, were unable to fully deal with the interacting local, national, regional, and global forces that have placed tremendous stresses on the continent and its peoples. The responses to these stresses, especially from the political elite, have been to resort to various forms of authoritarianism, force, and violence, which impacted nearly every part of the continent between 1960s to 1980s. These responses were however always countered by a quest for peace and stability as well as more inclusive and responsive forms of politics.

In the past three decades, in the aftermath of the collapse of the bipolar global world order and with the onset of the IMF and World Bank-led neoliberal structural adjustment policies, there have been considerable shifts in the landscape of conflict and peacebuilding in Africa. Violence proliferated across the continent in the 1990s and early 2000s (Adebajo and Rashid 2004; Williams 2016) as did conflict resolutions and peacemaking efforts. Much of that violence has ebbed. However, even as the different national, regional, and international actors have sought to constrain the use of violence as the primary instrument for the contestation and mediation of conflict, others have instrumentalized it as a legitimate strategy in the pursuit of economic, political, and particularist goals in Africa. There has, however, been the emergence of new terrains and actors in the political economy of violence, the proliferation of "structures of conflict," and increasing geopolitical complexity of so-called "local conflicts."

Consequently, Africa has been the site of extensive conflict resolution and peacebuilding interventions by local, national, continental, and international actors. The African Union (AU) and African regional economic communities, particularly the Economic Community of West African States (ECOWAS) and Intergovernmental Authority on Development (IGAD), have been engaged in peacekeeping. The AU declared 2010 as the "African Year of Peace and Security" and launched a "Silencing the Guns" campaign in February 2020. In the past 30 years, the largest number of UN peacekeeping missions (30 out of 72) have been deployed in Africa, and today the continent hosts seven of the 13 active UN peacekeeping contingents (Chapter 11). Special courts of justice, truth, and reconciliation commissions and extensive political-economic rebuilding programs have accompanied some of these programs. When the UN Peacebuilding Commission was inaugurated in 2005, its primary focus was on two African countries, Sierra Leone and Burundi, to which have now been added Liberia, Guinea, Guinea-Bissau, and the Central African Republic. Two historical political actors in Africa, namely former colonizers Britain and France, have independently and through the UN also been engaged in interventionist and peacebuilding activities in the continent.

At the same time, there has been a proliferation of institutions as well as research and publication on different aspects of conflict resolution, peacebuilding, and state building in Africa. Peace studies programs, institutes, and departments emerged in various African countries in the 1990s, and they continue to expand throughout the continent. Some of the peace studies programs are merely reproductions of similar programs in other parts of the world, often ill-adapted and inappropriate for African contexts. Similarly, even though the expanding body of scholarship in the field offers valuable insights on conflict prevention and resolution, post-conflict reconstruction, and state building in Africa, there is relatively little writing on epistemological and methodological approaches to guide emerging scholars on the nature and relevance of the conceptual tools and research strategies utilized in producing knowledge in the field.

The volume emerges out of these concerns and from a series of sustained scholarly dialogues about fieldwork preparation and research dissemination workshops for winners of individual research grants organized by the African Peacebuilding Network (APN) of the Social Science Research Council. These dialogues took place over the past five years in Botswana, Ghana, Kenya, Nigeria, Morocco, South Africa, Uganda, and Rwanda. They revolved around the epistemological, conceptual, and methodological gaps in the existing literature on peacebuilding. The participants also reflected on the struggles faced by many African scholars in refining the focus, methods, and research designs of their projects; in better aligning their research goals with their research methods; and sharpening their methodological skills before heading out into the field to collect data. The workshops also included questions and discussions about the extent to which the changing landscape of conflict, policy, and geopolitical shifts requires a rethinking of the use of concepts and their effects on theory and practice.

The rich, complex, and textured nature of the dialogues cannot fully be encapsulated in one volume. Yet, the spirit of dialogue, query, reflexivity, and attentiveness to the dynamic relationship between epistemology, methodology, and the field provides coherence across the different chapters. In reflecting on epistemology, there is the implicit agreement that conflict in Africa is not primordial, endemic, or irresolvable, but arises in the course of human interactions at multiple levels and in multiple domains. The chapters also underline the importance of sound and well-executed peacebuilding research. Such research must shed light on the root causes of conflict in Africa, the various mechanisms and regimes of violence, and the entanglements of politics and economy with other aspects of postcolonial life and therefore the possible conditions for durable peacebuilding. It must also be grounded in sound methodologies (Chapter 5 and 6), ones that are not simply borrowed wholesale, but carefully and critically calibrated to bring out the nuances and complexities of the different African terrains, voices, and experiences. Many of the chapters also approach the field not simply as a place of data collection, but as a site of active engagement with a range of conceptual framings in the study of peace and conflict in Africa, from standpoint feminist perspective to institutionalist to decolonial perspectives.

The dialogic relationship between theory, context, and the field is partly reflected in the organization of the book. The first section is mainly devoted to methodology, conceptual, and theoretical issues. The second section contains case studies illuminating field experiences while engaging substantive theoretical, methodological, and conceptual issues in peacebuilding research in Africa. What the contributions suggest is that it is not possible to isolate and compartmentalize the different aspects of peacebuilding research and practice. As Festus Aubyn (Chapter 1) specifies: "rather than a chronological step in a linear conflict management continuum, peacebuilding and the other elements of the peace process should be understood as interconnected, interdependent, and mutually reinforcing." By and large, most contributors concur on the fact that theory and methodology should not be seen as obstacles to a productive and reliable research endeavor, but as flexible tools of experimentation.

The landscape of peacebuilding research in Africa

Africa is neither on the margins of peacebuilding scholarship nor is peacebuilding no longer on the margins of knowledge production in the continent. The production of high-quality, high-impact, and policy-relevant peacebuilding research in and on Africa has been on the rise in the past few decades. This surge is partly due to the active support of think-tanks and research institutions of evidence-based research. These institutions include the UPEACE Africa Programme in Addis Ababa, the South African-based African Centre for the Constructive Resolution of Disputes (ACCORD) and the New York-based APN (see Omeje, Chapter 5 in this volume).

The surge in research has meant an active engagement and interrogation of the notion of peacebuilding. While the notion originated in the Western academy, there is no doubt that it gained considerable credence, especially amongst African researchers, with Boutros Boutros-Ghali's 1992 Agenda for Peace report. Boutros Boutros-Ghali's definition of peacebuilding in that report as a series of "actions to identify and support structures which strengthen and solidify peace in order to avoid a relapse into conflict" has been subjected to extensive discussion; its conceptualization and proposed practice of peacebuilding have gone through many changes as policy, experience, and research have come to interact in complex ways. Rather than the cohesion and the decisiveness that Boutros-Ghali had sought in responding to conflicts around the world, what had emerged in the field and scholarship was lack of consensus (Curtis and Dzinesa 2012; Karbo and Virk 2018; Omeje 2019). The absence of consensus on the scope, content, and purview of peacebuilding practices has implications for both the coherence of the field and the nature of interventions, as Norman Sempijja (Chapter 11) explains in this volume in analyzing successively An Agenda for Peace, the Brahimi report, and the Capstone doctrine.

Like many ideas and practices that have occupied a significant place in global imagination and politics, the evolution of peacebuilding has been accompanied by a growing body of literature from diverse geographic, disciplinary, and intellectual backgrounds and traditions. The literature, often contentious and sometimes complementary in its discourses, has attempted to make sense of conceptions, characteristics, activities, and consequences of peacebuilding activities. What has become clearly evident is that peacebuilding raises questions that cannot be adequately or fully tackled from one disciplinary perspective or scholarly tradition, but from interdisciplinary and intersectional perspectives. Many peace and conflict studies institutions and scholarship have in fact traditionally reflected such interdisciplinarity. Aubyn (Chapter 1) outlines the dominant themes that constitute peacebuilding research and practice on the continent. These include, but are not limited to,

> the concept of peacebuilding; the nexus between statebuilding and peacebuilding; the liberal peace debate; democratization and good governance; the root causes of conflicts; elections; conflict resolution and management; gender and peacebuilding; youth and peacebuilding; peacebuilding and peace operations; security sector reform (SSR); disarmament, demobilization and reintegration (DDR); transitional justice and reconciliation; economic recovery and growth; humanitarian relief; and regional dimensions of peacebuilding.

Yet interdisciplinarity does not always produce theoretical consistency. As Heidi Hudson (Chapter 2) argues, many critics of the liberal peacebuilding perspective have not yet provided strong enough and viable alternatives. For her, critical theories have not achieved the level of consistency and the neat fluidity between theory and practice that policy makers find so appealing in the liberal peace theory. On the other hand, Aubyn (Chapter 1) examines the specific absence of competing perspectives to the dominant universalistic

Western view in light of the neglect of so-called indigenous African perspectives and African worldviews. African scholars themselves have not produced anything qualitatively or epistemologically distinct to what has been produced elsewhere. Aubyn puts this down to African scholars' training in and dependence on Western scholarship.

These structural imbalances in the relationship and political economy of knowledge production and their effects on dominant narratives on African contexts are widely noted across the different contributions in this volume. In Chapter 12, Duncan Omanga and Pamela Mainye describe their peculiar journey with North American partners in a digital technology use related research in the context of electoral violence in Kenya. Despite growing possibilities for shifting narratives associated with innovative practices and technological creativity in commonly disadvantaged places, they point that some of these narratives can be manufactured to showcase "digital success" where a common donor-aid-logic continues to frame the outlook of technology users from a top-down approach. It is a relationship that ultimately constrains possibilities of theoretical and conceptual innovation on the part of African scholars.

This is not to say that African scholarship, especially narrative and theoretical innovation, is inexistent or invisible. In the past few years, African scholars have produced exciting new scholarship encompassing new perspectives on standpoint of feminist security perspective (see Chapter 2) to new institutionalist insights (see Chapter 3) to complex mixed methods analysis (see Chapter 5). Furthermore, Bouka, Berry, and Kamuru (2019), Peace Medie (2019), and Toni Haastrup (2019) have also recently produced fine-grained research on specific African conflict contexts and their entanglements with the broader processes of political and institutional change. These are examples of thoughtfully applied intersectional approaches to peace and security studies beyond the security agenda of the state.

Liberal peace and its discontents

Contemporary peacebuilding in Africa has been contentious and criticized because it is perceived as anchored primarily on a liberal peace perspective. The liberal peace perspective is inspired by a set of "standard interventions" that include conflict resolution strategies, demobilization, and reconstitution of security institutions, election-based democratization, market liberalization, and the restoration of civil and political rights (Paris and Sisk 2009; Omeje 2019). More than a post-conflict reconstruction framework, the liberal peace paradigm is an ideology that has modernist, civilizational, and colonial ambitions (Sabaratnam 2017). As Thomas Tieku (Chapter 3) contends, "liberal peacebuilding in contemporary Africa is a modernization or civilization project but disguised as neutral technical exercise designed to create durable peace in fragile states and to prevent post-war countries from relapsing into violent conflict." It is in that sense *a total normative project of transformation* of target societies. For many scholars, the argument of technicality and value-neutrality of the liberal peace framework is both misleading and unhelpful.

There is a degree of inconsistency that arises from a combination of normative prescriptions that come out of the liberal peacebuilding and an eclectic set of parallel theories. Notable advances in empirical and historical research have not been consistently matched by more sophisticated and coherent theoretical enquiries. While on the one hand the historical hegemony of the liberal peace theory has been firmly displaced, contending theories have not quite achieved a similar kind of transformative normative presence in the research field on the other. Hudson (Chapter 2) suggests breaking the pattern of unproductive binary opposition between postcolonial and decolonial theories by taking a multipronged critical feminist approach. In deploying the latter in dialogue with poststructuralist and postcolonial perspectives, scholars may be able to revive a critical tradition that brings into effect epistemological and methodological assessments in fundamental ways.

The emergence of what has been dubbed a second-generation peace research has coincided with growing interest in the intersection of conflict with resource governance and security, gender and transitional justice, development and reconstruction, and related questions (Paris and Sisk 2009; Karbo and Virk 2018). The enduring character of the liberal peace theory can be credited to its simplistic but coherent narrative that correlates a working society with functioning institutions. According to the liberal peace theory, states fail because institutions that safeguard individual rights and the free operation of the market have collapsed (Sogge 2011). A return to peace would therefore have to be achieved through state building. In addition to this apparent normative coherence, the liberal peace theory has served a useful instrument of international politics as it buttresses the legitimacy of a restricted number of Western states to arrogate themselves the role of norm creators and norm enforcers in the name of a putative international community. The more interventions are deployed on the basis of a universal attachment to a set of key values exemplified in the liberal peace framework, the harder it becomes to dismiss its underpinning assumptions as fallacious. However, as Amy Niang explains in her chapter, there are new types of conflicts best understood as complex new wars (CNW) and which are not appropriately explained by the liberal peacebuilding framework. According to Niang (Chapter 7), "these are conflicts that emerge at the confluence of a fraught, incomplete decolonization process, geopolitical struggles for control over strategic regions and their resources by powerful states, and post-Cold War reconfigurations of political regimes and global alignments."

Methodology, positionality, and language

Theories hold a strategic place in the research into conflict, the constitution of peacebuilding ideas, and the elaboration of policies that give validity to the latter. But theories, like policies, are never without purpose or certain preconceptions. The loaded character of theories is therefore to be constantly factored in research and policies as a potentially distorting variable for those keen to "apply" them as guiding lens. In short, theoretical engagement demands reflexivity. Reflexivity, Tieku (Chapter 3) tells us, should be exercised in two ways. It is as much about acknowledging the inevitably political place from which one is writing about and coming to terms with the fragility of any pursuit of objectivity. Scholars are not onlookers rubbernecking a field accidentally presented to their gaze. Instead, they are politically conscious data collectors who select "information" among various possible sources and in many different ways. Pretenses of detachment and objectivity should therefore be dropped in favor of a practice of reflexivity that keeps the researcher's many biases in check. As Fatma Ibnouf's (Chapter 8) thoughtful description of her experience studying caregiving in refugee camps in Darfur shows, reflexivity, like "agency" and "empathy" or "intent", is a practice that is deeply entangled in meaning-making exercises.

Reflexivity is more than a matter of awareness or disposition; it is also a necessary requirement in disentangling one's emotions, in unlearning received knowledge about a familiar or unfamiliar context, and in interrogating one's complex implication in a field of study. According to Maringira (Chapter 13): "when researchers study the people whom they had lived with, the research in itself reproduces 'home' and or 'familiar' identity." Just as the field reveals the tensions between the real and the imaginary, so can it unveil the ambiguities and contradictions associated with "insider" and "outsider" status. Jimam Lar (Chapter 9) shows in his research in Plateau, Bauchi, and Katsina states that his relative ease of access as an insider in his community is largely offset by suspicion from members of group involved in conflict with his. The in-group out-group dynamic can eventually constitute an impediment to exercising emotional restraint and a measure of distance from one's research object.

The complexities of location and positionality come up time and again in the various contributions as a question of perspective or vantage point and as a question of opportunity. Though researchers are composed of a bundle of social identities, they always think and write from a specific place and a specific set of values. Hudson (Chapter 2) states in this volume, "recognizing that our values as researchers shape all aspects of the research process is a fundamental step in the process of unlearning patriarchal and imperialist research practices which privilege the authority of the researcher." Nothing therefore in the way in which problems are identified and questions formulated is neutral or objective or entirely devoid of influence from the researcher's identity and values. Hudson further indicates, positionality itself is not an issue and can rather be "productively employed as a resource rather than an obstacle."

However, it is not enough to put a disclaimer on one's identity to fulfill the requirement of reflexivity. Reflexivity is an approach to the entire research and writing process and it needs to be constantly checked. It is about understanding the fact that researchers are never "outside" of the process of meaning-making and are rather involved in contextual power relations and that their subjectivities are heavily invested in the constitution of meaning and knowledge (Chapter 5). Specifically for a field that is evolving and consolidating into distinct schools of thought, reflexivity means at times an ability to assess the continued acuity of dominant frameworks against unexplored possibilities. Lar and Maringira (Chapters 9 and 13) offer instances of how researchers grapple with perceptions of their positionality in the field, and how they navigate insider positions that are at once assets and liabilities.

Methodology in peacebuilding research cannot be divorced from broader epistemological questions. Methodological tools have been adapted from various disciplines to respond to complex, changing theoretical and empirical imperatives. These tools have different provenances, levels of robustness, and degrees of efficacy. Rather than constituting an objective peace research standard, they are mainly reflections of researchers' strategic and therefore subjective choices. Omeje (Chapter 5) points out that "a weak understanding of qualitative research methodology and its contextual applications" is responsible for a misleading perception of qualitative methods as flexible and relatively unreliable and of quantitative methods as more rigorous and more reliable. Qualitative methods are as useful as they are adequately and rigorously applied to context following precise procedures of use. Whether one uses quantitative or qualitative methods of research, a complex approach to research objects can enable a researcher, according to Olayiwola Erinosho (Chapter 6), to uncover "the stories behind the stories." But this can only be achieved with the adequate use of adequate tools.

Understanding the power of language and discourse to constitute reality and to frame possibilities is also crucial in researching peacebuilding. The history of peacebuilding itself as a multilateral policy intervention is an example of the dynamic process of language and its (mis)uses. Concepts are political and ideological tools whose interactions with policies can be mutually reinforcing or distorting. In looking at the changing security order in the Sahel, Niang (Chapter 7) argues for instance that the War on Terror (WoT) has become an all-encompassing paradigm and a new epistemological line that dictates how we can study and understand the Sahel, its societies, and its security challenges. Humanitarian intervention in particular has been a prolific field of distortion and manipulation of ideologically loaded concepts such as the Responsibility to Protect. The American occupation of Iraq in 2003 and France and NATO-led intervention in Libya of 2011 are two cases in point. Whether peace is "built" or "managed" or "enforced" or whether an intervention is an occupation can be a matter of policy change or political manipulation. Furthermore, intervention can and does undermine ownership in thwarting the "self-organizing processes necessary to generate and sustain resilient social institutions" (Coning 2016: 166).

A recurring concern of the contributions in this collection is how to turn the specificity of African contexts into an opportunity to critically reassess theories, methods, and polices as they apply to peacebuilding research. For many, "the field" has become a productive site of unlearning and relearning. The authors concur that to confront one's deep-seated assumptions about "realities out there" in the contexts of research allows the researcher to dispel these, and to test methodology and research tools against "reality" and vice-versa. For "reality," according to Omanga and Mainye, can emerge out of a "complex tapestry of power hierarchies" to reproduce convenient and distorting narratives that obscure the distance and disconnect between the expectations of external researchers and institutions and the untamed complexity of contexts of conflict and its multiple actors.

Africa as a "Field" in peacebuilding

For the many scholars, consultants, analysts as well as international organizations like the United Nations, World Bank, and the IMF, the convulsions, conflicts, and peacebuilding efforts in Africa over the past three decades have made it the quintessential field site. Africa has become not only a site for theorizing about violence, the African "state" and African primordialities; it is a testing ground for various modes of containment and securitization; and it is a place of experimentation with various peacebuilding models. This perception of and disposition toward the continent and its people is not new; it is deeply rooted in the Euro-African entanglement over the past few centuries. As Ismail Rashid (Chapter 4) notes it is a disposition that has generated critique and resistance from several generations of African researchers and scholars.

Going to the Africa "field," which fiercely refuses generalization, demands reflexivity on the part of researchers of peacebuilding. Often in situations where researchers have found themselves unable to prove their theories, they have either found research objects/targets inadequate or their "field" to be unsuitable for their tools rather than confront the latter's limitations in explaining the reality under scrutiny. In light of this, "the field" has to be reconceptualized as a competing school for learners on a par with formal methodology courses. The field is the place where theories are confronted to the contradictions of context and their inconsistencies. As Asebe Regassa Debelo (Chapter 10) puts it, fields of conflict are complex webs of "culturally embedded practice that is performed within broader ontological and epistemological understandings of people."

In the African "field," these broader ontological and epistemological understandings must necessarily be embedded within a critical ethical framework. Ethics in research is not simply a box-ticking exercise in research approval forms; it is ultimately about safeguarding the integrity of the entire research process. Rashid, Ibnouf, and Erinosho emphasize in this volume that ethical considerations should permeate researchers' choice of subject, research questions, conceptual approach, methodology, and field research locations. Africa's history makes the avoidance of unnecessary harm for all involved in a research project more than imperative. Researchers have to be conscious of respecting the dignity of research subjects, especially those who are vulnerable and marginalized, and behave justly, honestly, and beneficently in the field. Above all, the different contributions in this volume underline that to rethink research ethics is to rethink the researcher/researched relationship, and therefore the subject/object framework, with its implied power structure, knowledge hierarchy, and extractive logic.

The rethinking of the ethics of peacebuilding and fieldwork requires that researchers become the learners, and be conscious that they are the ones who stand to benefit from immersion into known and unknown fields of experience. It requires therefore that the subject/object relationship be turned into a subject/subject relationship. It bears re-emphasizing the inherently intersubjective and dynamic nature of the field, and the need to constantly interrogate the cultivated, rhetorical distance between researchers and researched. Humility and empathy are thus qualities as valuable as curiosity and openness (i.e. from a place of unknowing). The African field pulsates with possibilities only available to the humane, the curious, and the deeply reflexive researcher.

Some of the case studies in the volume allude to some of these possibilities, especially in some of their methodological strategies and conceptual approaches to understand particular local and cultural dimensions of peacebuilding (Ibnouf and Debelo in this volume). Ibnouf's field immersion (Chapter 8) in the everyday experience of women caregivers in Darfur is imbued with research ethics that is not only rigorous but also humane and dialogical. In Ibnouf's analysis, women constitute and sustain forms of social organization during and after conflict. The roles they perform in extremely precarious conditions are not captured in a victim/perpetrator binary analysis but compel the researcher to explore a number of invisibilized dimensions in conflict contexts. In doing so, Ibnouf is able to center both the sources and the dynamics of knowledge around the experiences of so-called marginalized women.

Conclusion

There is no indication that even if the guns were to become silent in Africa this year as advocated by the AU and supported by the United Nations that peacebuilding research in the continent would slow down considerably. This volume contributes to the ongoing process of research in three ways. First, it brings together three aspects of the study of peacebuilding in Africa, namely the methodological, the theoretical, and the practical (both in terms of empirical and policy relevance). It provides a toolbox of perspectives on conceptualizing and doing peacebuilding research in Africa that are valuable to research, policy, and practice in African contexts and more broadly. Second, it rethinks the implications, for research strategies and for knowledge production, of the shifts noted earlier alongside urgent questions that challenge conventional approaches to peacebuilding research on the continent.

These questions range from the requirement for paradigmatic shifts to the emancipatory potential of engaging research subjects as actors who generate specialized knowledge. Third, the volume adopts a perspective *from* Africa as a way of connecting context to genres of knowledge to a broader potential of research to support ongoing initiatives and programs to build sustainable peace on the African continent. It showcases projects that are informed by years of engagement with the field and whose insights can be used to inform thinking, research, the policy, and practice of peacebuilding in Africa.

References

Adebajo, Adekeye and Rashid, Ismail. (2004) *West African Security Challenges: Building Peace in a Trouble Region*. Boulder: Lynne Rienner.

Bouka, Yolande, Berry, Marie E. and Kamuru, Marilyn Muthoni. (2019) "Barriers and Enablers of Women's Political Participation in Kenya's Devolved Political System," *Journal of Eastern African Studies* 13(2), pp. 313–333.

Coning, Cedric. (2016) "From Peacebuilding to Sustaining Peace: Implications of Complexity for Resilience and Sustainability," *Resilience: International Policies, Practices and Discourses* 4(3), pp. 166–181.

Curtis, Devon and Dzinesa, Gwinyayi. (2012) *Peacebuilding, Power and Politics in Africa*. Athens: Ohio University Press.

Haastrup, Toni. (2019) "WPS and the African Union," in True, J. and Davies, S. (eds.) *The Oxford Handbook on Women Peace and Security*. Oxford: Oxford University Press, pp. 375–387.

Karbo, Tony and Virk, Kudrat. (2018) *The Palgrave Handbook of Peacebuilding in Africa*. Cham, Switzerland: Palgrave Macmillan.

Medie, Peace. (2019) *Global Norms and Local Action: The Campaigns to End Violence against Women in Africa*. New York: Oxford University Press.

Omeje, Kenneth. (2019) *Peacebuilding in Contemporary Africa: In Search of Alternative Strategies*. London and New York: Routledge.

Paris, Roland and Sisk, Timothy. (2009) "Introduction. Understanding the Contradictions of Postwar Statebuilding," in Paris, Roland and Timothy D. Sisk (eds.) *The Dilemmas of Statebuilding. Confronting the Contradictions of Postwar Peace Operations*. London: Routledge, pp. 1–20.

Sabaratnam, Meera. (2017). *Decolonising Intervention. International Statebuilding in Mozambique*. London: Rowman & Littlefield.

Sogge, David. (2011) "Weak States and the Savage Wars of Peace" *Global Dialogue* 13(1), Winter/Spring 2011, pp. 6–18.

Williams, Paul. (2016) *War and Conflict in Africa*, Second Edition. Cambridge: Polity Press.

Concepts, theories and methodologies

1 An overview of recent trends in African scholarly writing on peacebuilding[1]

Festus Kofi Aubyn

Introduction

Since the Agenda for Peace report in 1992 and subsequent reports, including the 2000 "Brahimi Report" and the 2004 report of the High-Level Panel on Threats, Challenges and Change, the meaning of the term "peacebuilding" has expanded significantly (United Nations 1992, 2000, 2004). From Boutros Boutros-Ghali's original definition, "actions to identify and support structures which strengthen and solidify peace in order to avoid a relapse into conflict," peacebuilding is today understood to encompass the full array of processes, approaches, and stages needed to transform conflict situations into more sustainable peaceful relationships and development before and after conflicts (Lederach 1997; Adetula, Bereketeab and Jaiyebo 2016). The expansion of the term has been accompanied by a growing body of literature from diverse geographic, disciplinary, and intellectual backgrounds and traditions that have attempted to make sense of peacebuilding activities and their consequences. Ismail Rashid points out that in Africa, the proliferation of conflicts has made researching and writing on peacebuilding a necessity (2017). Ever since the term peacebuilding rose to prominence in 1992, various African institutions, practitioners, and scholars from different disciplines, such as political science, history, anthropology, sociology, international relations, development studies, and peace, conflict, and security studies, have written on different facets of peacebuilding from diverse perspectives and contexts. African scholarship has grown over the years in parallel with the expansion in peacebuilding activities and the proliferation of institutions at the global, regional, and national levels tasked with peacebuilding work (Curtis and Dzinesa 2013).

But what have been the key trends, themes, and debates in the African peacebuilding literature? How has African peacebuilding literature contributed to the current global discourse, policy development, new knowledge, and practice? And to what extent has African thinking shaped the global peacebuilding agenda, especially in Africa? These questions are particularly critical due to the popular impression that peacebuilding in Africa is determined in Western capitals and in the corridors of the UN, which neglects how local ideas shape the global policy debates and practices (ibid., p. 3). As David J. Francis rightly puts it, the field of peace studies has been crafted in Euro-American institutions, largely by scholars from Europe and North America, neglecting the fact that Africa had well-developed and tested indigenous approaches and methods of conflict management and peacebuilding. This universalization of peacebuilding theories and methodologies has led to a particularistic and dominant discursive hegemony of Western explanations of what is considered peace, thereby limiting the "Africanist worldview" of the term peacebuilding (2007). Highlighting the diverse ideas and views of African scholars, practitioners, and institutions is therefore imperative to deepen understanding of the impacts and limitations of recent peacebuilding interventions in Africa.

Against this backdrop, this chapter identifies and reviews some of the major academic writings by African scholars and institutions on peacebuilding since 2007. It begins by tracing the evolution of the growing body of African peacebuilding scholarship and then analyzes the key trends, themes, and debates in the African scholarly writings and their connections or disconnection with global discourses and processes. The next section discusses the strengths and weaknesses of the peacebuilding literature produced by African scholars, including their contributions to the global peacebuilding scholarship.

Tracing the evolution of peacebuilding scholarship

As a field of study, peacebuilding has evolved over the past three decades primarily as a subfield of peace and conflict studies in the social sciences. The intellectual antecedent of the concept can be traced to the peace philosopher and researcher, Johan Galtung, who coined the term in his pioneering paper "Three Approaches to Peace: Peacekeeping, Peacemaking, and Peacebuilding" (1976). It was after this seminal essay and subsequent work by John Paul Lederach that the term began to gain currency in the academic world (1997). However, most of the initial peacebuilding research at the time was dominated by Western scholars from peace research institutions and universities in Europe and North America. The peacebuilding scholarship in Africa during the period had not yet developed, although Erin McCandless and Tony Karbo posit that some elements of the field had developed for decades in the social sciences in Africa (2011). This is because most of the traditional practices to resolve conflict and to promote social harmony in Africa that historians, anthropologists, and sociologists, for example, did research on were not described using the term "peacebuilding." Equally important is what scholars describe as the state of affairs in the 1960s and 1970s, when most post-colonial government considered peace research a subversive activity; this inhibited the development of local research on the subject (McCandless, Bangura, King and Sall 2007; McCandless 2011). According to Eghosa Osaghae, local and foreign researchers were critiqued for giving intellectual muscle to a supposedly neo-colonial project meant to keep the fragile African states weak and divided (2001, p. 119).

Nonetheless, the situation began to change in the aftermath of Boutros Boutros-Ghali's 1992 landmark report, Agenda for Peace, which introduced peacebuilding into UN vocabulary more systematically. Ghali's report stimulated academic interest and research among African scholars both within and outside the continent on conflict resolution, peacekeeping, and peacebuilding in an attempt to find lasting solutions to the intractable conflicts in Africa. Since then, peacebuilding research has emerged as a recognized domain of rigorous study of peace and conflict in Africa. However, it is important to stress that the evolution of the academic writing on peacebuilding from the 1990s to date has been influenced by certain key developments on the continent.

To begin with, the increasing number of complex intra-state armed conflicts in the early post-Cold War period and the resultant need to develop coherent robust responses to promote sustainable peace was one such development (Jaye 2009). These protracted armed conflicts attracted academic and policy interest on how to build sustainable peace and development to prevent countries from relapsing into conflicts after their initial resolution. Although such interest was more pronounced at the policy level, African academics in and outside the continent began to analyze some of the peacekeeping and humanitarian interventions from theoretical, methodological, and empirical perspectives. Many of the scholars and practitioners focused their research on analyzing the root causes of conflict in countries such as Liberia, Sierra Leone, Rwanda, Somalia, Burundi, Guinea Bissau, and Democratic Republic of Congo (DRC), and how to build lasting peace and human development in the post-conflict context (Anyidoho 1997; Aboagye 1999; Francis 2000; Jaye 2003; Adebajo 2004; Adebajo and Rashid 2004; Olonisakin 2008).

Others also analyzed the interventions of inter-governmental organizations, such as the Economic Community of West African States (ECOWAS), the Organization of African Unity (OAU)—now African Union (AU)—and the UN in some of the war-torn countries. Francis, for instance, discussed the experiences of ECOWAS in West Africa, and asserted that regional organizations were the most viable framework for the maintenance of regional peace and security, contrary to the popular notion that they play very limited role. Similarly, using Liberia as a case study, Comfort Ero argued that despite the numerous challenges and setbacks, ECOWAS interventions in West Africa were a major step in the maintenance of regional peace and security. Many other scholars also provided case studies and empirical perspectives on the causes of conflicts and peace processes in Liberia, Rwanda, Sudan, Burundi, and Sierra Leone (Ero 1995; Anyidoho 1997; Jaye 2003; Aboagye 2004; Olonisakin 2008; Francis 2009; Obi 2009, p. 119; Bah 2010). Nevertheless, some of this literature was critical of the interventions and state rebuilding processes in post-conflict countries by the UN and other external actors, which sometimes resulted in further instability and violence (Nwolise 1992; Aning 1997; Adebajo 2002; Albert 2008; Malan 2008).

Another major development was the growth of peace research centers and institutes across the continent and the introduction of peace and conflict studies programs in some African universities (Omeje, 2015b). Even though the education system in Africa was slow to respond to the demands for peace studies in the 1980s, the mid-1990 onwards saw a vast expansion in the number of peace-related professional bodies and research institutes, the proliferation of university courses, and the launch of academic journals on conflict resolution, management, and peacebuilding in Africa (Omeje 2015a, 2015b). Mccandless and Karbo mention the Council for the Development of Social Science Research in Africa as one of the first major centers in the 1990s for the study of peace and conflict on the continent (McCandless and Karbo 2011). According to Isaac O. Albert, universities in Ghana, Kenya, Liberia, Nigeria, Sierra Leone, South Africa, and Uganda also began offering academic programs in peace studies which treated peacebuilding as a peripheral subject (2017). Some of these universities (with the date the academic programs were established) are: The University of Ibadan (2000) and the University of Ilorin in Nigeria (2008), the University of Cape Coast in Ghana (2007), Makerere University in Uganda, and Kenyatta University in Kenya (2010). A plethora of think tanks, civil society organizations (CSOs), and research institutions were also established as centers for training and research on conflict resolution, management, and peacebuilding. These include the African Centre for the Constructive Resolution of Disputes (ACCORD 1992), the Centre for Conflict Resolution (CCR 1994), Institute for Peace and Security Studies at Addis Ababa University (IPSS 2007), Cairo International Centre for Conflict, Peacekeeping, and Peacebuilding (CCCPA 1994), the Institute for Security Studies (ISS 1991) in South Africa, the Kofi Annan International Peacekeeping Training Centre (2004), the West African Network for Peacebuilding (WANEP 1998) in Ghana, and the Nairobi Peace Institute (NPI 1984) in Kenya.

The establishment of such academic programs and research institutions was accompanied by the launch of various peer-reviewed journals on peace and security, conflict resolution, and peacebuilding in Africa. Among these were the *Africa Peace and Conflict Journal* (APCJ), launched in 2008 by the University for Peace in Ethiopia; the *African Journal on Conflict Resolution* established by ACCORD in 1999; and the *African Security Review* established by ISS in 1992. These journals encouraged the publication of high-quality, innovative work by African authors on a wide range of peacebuilding issues in Africa and provided a forum for the exchange of critical thinking emerging from African universities, colleges, and organizations. They also contributed to the development of ideas, writings, and dialogue in the field of conflict management and peacebuilding on the African continent.

Over time, academic interest and research on various aspects of peacebuilding has become widespread on the continent, particularly after the establishment of the UN Peacebuilding Commission in 2005, given that all the countries on the Commission's agenda are from Africa (Aning and Lartey 2010).[2] The development of normative instruments at the regional level also generated interest in the subject, with few studies focusing on how the AU can effectively implement its post-conflict reconstruction framework.

Key themes and perspectives on peacebuilding

From the existing literature, one can discern a number of trends and perspectives in the analysis of peacebuilding in Africa. Some of the key trends and themes that have dominated the literature in the past decade include the concept of peacebuilding; the nexus between statebuilding and peacebuilding; the liberal peace debate; democratization and good governance; the root causes of conflicts; elections; conflict resolution and management; gender and peacebuilding; youth and peacebuilding; peacebuilding and peace operations; security sector reform (SSR); disarmament, demobilization and reintegration (DDR); transitional justice and reconciliation; economic recovery and growth; humanitarian relief; and regional dimensions of peacebuilding. Some of these trends and themes are examined below.

The concept of peacebuilding

Peacebuilding involves the full range of initiatives, measures, and activities aimed at preventing, reducing, and transforming conflicts and developing institutions, attitudes, and relationships that promote sustainable peace and development. Tim Murithi describes it as a process whose goal is to strengthen the capacity of societies to promote a "positive peace" (2006). This involves activities aimed at addressing the root causes of conflict, prevention, and mitigation of all forms of violence, and the promotion of reconciliation and coexistence on the basis of human rights and social, economic, and political justice.[3] Based on the earlier definitions, peacebuilding can be understood as a set of activities that promote both positive and "negative peace" (absence of violence). This understanding of peacebuilding differs from the initial conceptualization of the term by Boutros-Ghali in his widely cited report, An Agenda for Peace, and the supplement to the document published in 1995. In An Agenda for Peace, peacebuilding was defined as "actions to identify and support structures which strengthen and solidify peace in order to avoid a relapse into conflict."[4] Peacebuilding was conceptualized chronologically; preventive diplomacy, peacemaking, peacekeeping, and peacebuilding were considered separate phases of the peace process. As Cedric de Coning posits, conflict prevention and peacebuilding were "juxtaposed at the opposite ends of the conflict management spectrum, with preventive diplomacy representing the first or opening stage of an intervention." If unsuccessful, this would be followed by peacemaking; if an agreement is reached between conflict parties, this could result in a peacekeeping mission. Finally, peacebuilding would ensure that the conflict does not reoccur by addressing the root causes.

In the past two decades, African researchers and institutions, like their counterparts elsewhere across the globe, have through research and practice demonstrated that the transition from war to peace rarely follows a linear path as suggested by Boutros-Ghali's report. Siphamandla Zondi notes that the AU's Post-Conflict Reconstruction Policy (PCRD) framework maintains that there is no need to place these activities in a sequence because it does not accept the logic that one element must be fully in place before the next phase kicks in (2017). Consequently, in the 2015 review of the UN peacebuilding architecture, the concept of "sustaining peace" was introduced to liberate peacebuilding from being strictly limited to post-conflict contexts. Activities to build peace are now understood to take place before conflict and not only limited to the post-conflict context. That is, peace and war may exist simultaneously in different parts of the same country as in the case of Sudan, where conflict escalated in Darfur even after the Comprehensive Peace Agreement (CPA) formally ended the conflict between the North and South in 2005. Another typical case is Uganda, which is often described as a "peaceful" country, but has had a long-lasting conflict with the Lord's Resistance Army in the north.

Rather than a chronological step in a linear conflict management continuum, peacebuilding and the other elements of the peace process should be understood as interconnected, interdependent, and mutually reinforcing (de Coning 2010, p. 12). Albert summarizes this sentiment by indicating that the term "peacebuilding" has gone through such a profound transformation that it today encapsulates all the other peace terms promoted in the Agenda for Peace report (2017, p. 5).

The liberal peace paradigm

The liberal peace paradigm, which is linked to the wider debate on democratic peace theory, has been a major premise of internationally supported peacebuilding efforts in Africa since the early post-Cold War era. As a paradigm derived from a long tradition of Western liberal theory and practice, liberal peace is about building peace within states on the basis of liberal democracy and market economics (Lemay-Hebert 2013). Though the main components of the liberal peace thesis vary, it usually includes the promotion of multiparty electoral democracy, constitutionalism, rule of law, good governance, human rights, SSR, civil society promotion, gender equality, economic reforms, and privatization (Salih 2017). These political and economic liberalization programs, previously regarded as effective antidotes to violent conflicts in war-torn societies, have been deeply questioned. Questions about whether the promotion of multiparty electoral democracy and institution building, legal reforms, and market- based economic reforms actually lead to sustainable peace or whether they undermine the consolidation of peace in post-conflict societies abound in the literature (Zambakari 2016; Bah 2017). While some African researchers have written extensively on different variants of the liberal peace thesis, others have challenged or critiqued the liberal premise of contemporary peacebuilding interventions.

One of the important issues that has generated a lot of debate in the literature concerns the appropriateness of the top-down approach (state-centric) of liberal strategies in dealing with the needs of post-conflict countries. Contributing to this debate, Coning argues that for a peace process to become sustainable, "resilient social institutions need to emerge from within, i.e. informed by the local culture, history, and socio-economic context," instead of the liberal top-down problem-solving approaches aimed at resolving conflict. International actors, he noted, can assist and facilitate that process, but if they interfere too much they will undermine the "self-organizing processes necessary to generate and sustain resilient social institutions" (de Coning 2016). In the same way, Funmi Olonisakin and Alfred Muteru in their article, "Reframing Narratives of Peacebuilding and Statebuilding in Africa," questioned the current top-down approach to peace that is based on the assumption that a certain type of state (democratic) would lead to peace and stability. To them, the top-down approach often ignores local contexts, informal actors, and initiatives, which if brought on board could lead to longer-term, sustainable, context-specific programs that produce better outcomes (2014). Citing some practical cases to support this point, Josaphat Musamba Bussy and Carol Jean Gallo locate the cause of the failures of the UN and World Bank peacebuilding interventions in Eastern DRC within the liberal governance framework. According to them, these interventions failed because they were based on the false premise that the absence of a strong liberal state is the primary reason conflict persists in the DRC, leading to the implementation of programs in a primarily top-down fashion, despite discourse on the importance of local ownership (2016).

Indeed, the focus on the local context in peacebuilding interventions is seen by most African researchers as a critical factor in efforts to build peace, as one-size-fits-all solutions do not work in all post-conflict situations. However, internationally supported peacebuilding interventions in Africa often overlook the local community's conception of peace, justice, and order, which is rooted in their local culture and traditions.[33] The top-down approach to peacebuilding not only downplays the contributions that African traditional systems make toward ensuring peace, but sometimes perpetuates conflict situations. These concerns have led some African researchers and practitioners to view liberal peacebuilding as simply a hegemonic enterprise forced on post-conflict societies in Africa to promote the economic and political interests of the Western world. As Albert rightly puts it, "it is an external imposition that is not embedded in African culture and experiences" (Albert 2008, p. 32). However, there are authors like Abu Bakarr Bah who although doubtful of the feasibility of the liberal approach advance the notion of people-centered liberalism as an approach that can lead to stable, democratic, and economically successful post-conflict states. The notion of people-centered liberalism taps into the neoliberal value of freedom and the principles of economic and social well-being implicit in the developmental state. His proposal is premised on the fact that experimentation with unorthodox variants of liberalism in Africa since independence has yielded limited results, hence the need for a people-centered liberalism (Bah 2015). Other scholars have also suggested an approach based on resilience, human security, and hybridization processes as a way to capture the complexity of the interactions between internal and external actors in peacebuilding contexts (Aning and Aubyn 2018). It is instructive, however, to note that "liberal peace" is the overarching idea that has influenced the dominant themes discussed in the subsequent sections of this chapter, which include statebuilding, women in peacebuilding, youth and peacebuilding, economic recovery and growth, SSR, and transitional justice and reconciliation.

Socioeconomic development and peacebuilding

Conflicts in Africa have often led to the destruction of economies, the social fabric of societies, infrastructure, and educational and health services in the affected countries. Promoting socioeconomic development in the aftermath of conflict is therefore necessary not only to ensure economic growth, trade, investments, and human development but also for the consolidation of peace. In this regard, the AU stresses the need for socioeconomic development to improve the living conditions of people by meeting their basic needs, such as health, education, and food, and by reducing poverty and inequality through the creation of wealth, livelihoods, and promoting equal access to resources (African Union 2006). Given the challenges and particularities of post-conflict settings in Africa, the literature provides different perspectives about which models and approaches should inform socioeconomic development, how economic recovery efforts should be funded to secure peace, and the role of socioeconomic factors in creating or exacerbating conflicts.

One issue that has generated discussion among African scholars is the question of economic liberalization, which emerged from the liberal peace paradigm. In general terms, the liberal peace approach is often criticized for its inability to adequately address the specific socioeconomic difficulties of post-conflict societies, thereby exacerbating social and economic inequalities through marginalization and increased vulnerability to poverty. Critics, such as Mohamed A. Salih, argue that the liberal peace paradigm has largely failed to deliver tangible development or economic benefits to the majority of the African poor (2017). He attributes this failure to the dominant political economy of the liberal peace framework which has proven unable to address major developmental problems such as poverty, exclusion, social justice deficits, and inadequate access to basic human needs. In countries such as Angola, Liberia, and Cote d'Ivoire, Devon Curtis and Gwinyayi A. Dzinesa suggest that rapid liberalization strategies, efforts to hold quick multiparty elections, and rushed economic reforms have had destabilizing results instead of creating the conditions for stable and lasting peace (Curtis and Dzinesa 2013). In Liberia, for example, Cyril Obi, along with Thomas Jaye and Tobias von Gienanth, maintains that Charles Taylor, who won the post-conflict election in 1997 after "successful" peacebuilding efforts, dismantled the democratic elements of the state and repressed his political opponents, triggering a relapse into conflict in 2003. Furthermore, in Angola, Eritrea, Mozambique, Namibia, Rwanda, Sierra Leone, and South Africa, it has been argued that the outcome of the transition from conflict to peace and development was not motivated by the uncritical superimposition of liberal peace but rather the political economy of peace. In this respect, Peter Kagwanja was probably right when he claimed that liberal peacebuilding is "better in ending wars" than in fixing democracies (2017).

Statebuilding and peacebuilding

The relationship between statebuilding and peacebuilding in Africa is another issue that has attracted both academic and policy attention. While some scholars conceive peacebuilding as a subset of statebuilding activities, others consider statebuilding to be part of the peacebuilding process. The latter view is shared by many African researchers and institutions. Karbo, for example, states that statebuilding is vital to peacebuilding because both concepts seek to address the root causes and effects of conflict by restoring broken relationships between the state and society, building state institutions, and undertaking political reforms (Karbo 2012). Moreover, both concepts have the mutually reinforcing goals of promoting peace in fragile and conflict-affected states and strengthening state-society relations to promote inclusive political systems. Statebuilding and peacebuilding have become very important for Africa's stability and development because since independence, most African countries have been characterized by intra-state conflicts, political instability, separatist rebellions, and underdevelopment (Murithi 2008, pp. 70–82; Okyere 2015). Despite remarkable progress during the past decade, Africa continues to face serious challenges that sometimes threaten the very foundations of states. Dele Olowu and Paulos Chanie assert that the fragility of the African state can be traced to colonial history; most states created after 1950 from the ashes of colonial empires were weak and unable to fulfill the basic functions of the state in a Weberian sense (2016). According to Murithi, this externally imposed problem of the nation-state in Africa is exacerbated by internal factors, such as bad leadership, corruption, and adversarial politics, leading to violent conflict situations on the African continent. This has created situations of fragile social peace and led to the destruction of the basic elements of political order that promote structural stability and resilience. Consequently, there is a mismatch between the demands of the citizens and the capacity of the state to adequately fulfill those demands, resulting in state fragility (Murithi 2012).

In this context, statebuilding is constructed as a prerequisite to restoring and rebuilding the institutions and apparatus of fragile states for effective service delivery in order to consolidate peace and contribute to long-term stability. Without undertaking a comprehensive statebuilding process—especially in the fields of political governance, SSR, economic reforms, and social justice—Kwesi Aning and Naila Salihu contend that there is a greater risk that the country concerned will revert to conflict (2012). However, the international statebuilding practices in Africa face some limitations. Some scholars have questioned the liberal approach underlying statebuilding processes on the basis that building or reconstructing the state does not necessarily constitute the clearest pathway to peace. While the idea of supporting post-conflict societies to pursue their statebuilding efforts seems logical, the liberal peace assumption underpinning it that one template will suit all conflict situations is faulty at best. According to these scholars, the one-size-fits-all approach to post-conflict states transitioning from war to peace based on itemized pillars and timelines only leads to negative peace and not positive peace (Olonisakin, Ababu and Muteru 2017). Likewise, others have argued that the externally driven statebuilding processes in Sierra Leone were not designed to support stable peace. To them, the institutional statebuilding approach, with its emphasis on institutional efficiency, technocratic support, macroeconomic stability, and consolidation of state authority, ignored the historical and sociological process in the country's state formation processes (Nyokabi, Ipke and Alao 2014).

Similarly, Christopher Zambakari blames the failure of statebuilding efforts in South Sudan to sustain peace on what he describes as the misguided liberal program implemented shortly after the signing of the CPA. This failure, he maintains, stems from the prioritization of statebuilding by the donor community at the expense of nation-building activities which, according to Mahmood Mamdani, could have fostered social cohesion within and between communities and dealt with the legacies of war. This brings to light the quandary of whether peacebuilding, as it is currently practiced in most post-conflict countries, is capable of providing a basis for statebuilding and inclusive political systems in Africa. Instead of imposing what have been described as flawed assumptions, external actors must exercise great caution, sensitivity, and flexibility to ensure that they do no harm through their interventions (Mamdani 2016; Zambakari 2016).

Security sector reform

SSR has become central to conflict prevention, stabilization, peacebuilding, and development agendas across Africa. The concept emerged in the 1990s as part of an international agenda to promote sustainable peace and development in societies transitioning from conflict or long-term authoritarian rule. It generally covers those institutions entrusted with the protection of the state and its citizens, such as the military, paramilitary forces, and intelligence services; civil authorities that provide oversight of the security sector, such as parliament and the ministries of defense, finance, and interior; and the justice and law enforcement institutions, such as the judiciary, police, penal services, and traditional justice systems (Adedeji 2007a; Bryden, N'Diaye and Olonisakin 2008).

The literature points to multiple reasons why the implementation of SSR in post-conflict contexts in Africa is of strategic importance. Some scholars claim that most security agencies in Africa after independence were solely established for the purpose of regime security rather than providing security for citizens (Adedeji 2007b, 2007c; Jaye 2008; Malan 2008; Gbla 2007). As a result, these institutions have become instruments for the repression of political opposition elements, objects of fear, and are often implicated in the perpetration of atrocities and human rights abuses. This particular historical trajectory accounts for some of the present weaknesses in the security sector, making reform necessary Within this context, it is clear why scholars have put great emphasis on defining African security within the human security paradigm with the goal of placing individuals, rather than the state, at the heart of security considerations. Some emphasize the importance of other SSR priorities, including the problem of factionalism and fragmentation within the security sector after armed conflicts, the blurring of roles and responsibilities between the various elements of the sector, and bringing the security sector under the regulation of democratic institutions (Adedeji, 2008).

At the heart of the SSR debate are the questions of what should constitute SSR and which institutions should be involved in the process. For example, should reform encompass both state-level institutions and non-state actors or only the former? While most donor agencies and development partners focus on state-level actors, many African researchers argue for the inclusion of both state and non-state actors in the SSR process. The African perspective is premised on the fact that in many African contexts the state never holds a monopoly on the legitimate means of coercive violence, unlike in Max Weber's conceptualization (Aning and Aubyn 2018).

Across the continent, the state is increasingly being challenged by non-state actors such as armed militias, private security companies, vigilante groups, terrorist groups, and multinational corporations. This calls for a paradigm shift away from the Westphalian, state-centric notion of SSR to a more holistic understanding of the security sector to ensure that interventions respond to the dynamics of the relevant environments. Related to this argument is the issue of how SSR should be conceived, either narrowly defined focusing on reform in specific sectors (defense and police reform) as in the case of Liberia or comprehensive, covering all the key institutions of state as implemented in Sierra Leone (Von Gienanth and Jaye 2007). Many scholars argue for a comprehensive approach that encompasses not only the military and police institutions but also intelligence agencies and the criminal justice system. The narrow focus of SSR on the military and the police, as was the case in Liberia, is often criticized by those who claim it weakens other elements of the security sector (Adedeji 2007b; Malan 2008; Jaye 2009).

Another interesting debate centers on whether the overarching goal of SSR should be reform or transformation of the security sector. African institutions like the ECOWAS argue for security sector transformation rather than reform (Adetula, Redie, and Olugbemi 2016). Two main arguments are advanced in the literature in support of this strategy. The first is that transformation will often improve the strained relations between government, civil society, and the security establishment. The second is that it leads to a complete overhaul of the structures and institutional cultures of security organizations and their relations with civil authorities toward alignment with democratic governance principles (Adedeji 2007b; Bryden, N'Diaye and Olonisakin 2008; Jaye, Garuba and Amadi 2011). This sharply contrasts with reform processes which gradually promote democratic civil-military relations without addressing the underlying institutional flaws that entrench patterns of abuse and insecurity in Africa.

Transitional justice and reconciliation

According to the AU, the purpose of transitional justice frameworks is to address the legacies of large-scale past abuses to ensure accountability, promote justice, and achieve peace and reconciliation (African Union 2013).[5] A review of the literature shows the implementation of a wide range of transitional justice approaches ranging from judicial mechanisms, including international criminal tribunals, hybrid courts, and domestic courts to non-judicial mechanisms like truth commissions, reparations, and tradition-based processes. Many of these studies focused on the processes, mechanisms, and limitations of the transitional justice frameworks. The relevant examples often cited in the empirical studies reviewed include retributive and restorative justice mechanisms such as the Gacaca courts in Rwanda; the Palava Hut in Liberia; Amnesty; Truth and Reconciliation Commissions in South Africa, Liberia, Mali, and Kenya; and the international criminal tribunals established in Rwanda and Sierra Leone, in addition to the International Criminal Court (ICC) (James-Allen, Weah and Goodfriend 2010; Aning and Jaye 2011; Kariuki 2014; Danso 2016).

Recent cases that have received considerable attention in academic and policy circles are the prosecution and trial of African leaders like Charles Taylor of Liberia and Laurent Gbagbo of Ivory Coast by the ICC as well as the charges of alleged crimes against humanity during the 2007–2008 post-election violence in Kenya against William Ruto and Uhuru Kenyatta, which were later terminated.[6] The charges and trials against these leaders, according to a 2013 report by the African Union Panel of the Wise, have led to increasing dissent in Africa due to the perception of threats to state sovereignty, the intrusiveness of international legal institutions toward weak states, and the concerns about the selective application of justice (African Union of the Wise 2013).

Generally, the key issue that has occupied most of the studies is the relative efficacy of the different transitional justice models and how they contribute toward the restoration of peace, social harmony, and reconciliation in post-conflict societies. While some scholars think that transitional justice should be pursued through punitive or retributive justice models such as the international criminal tribunals and the ICC, others have proposed a range of restorative justice and reparation models such as truth commissions and customary justice systems. Advocates of the retributive justice model claim that it contributes toward reconciliation, reaffirms the rule of law, makes amends to victims by holding perpetrators accountable for past wrongdoings, and serves as a deterrent to future abuses by signaling the boundaries of socially acceptable behavior (ibid.). Critics of this approach doubt that peace can be consolidated in divided societies through the criminal justice system. They question whether this legalistic approach leads to sustainable peace, especially in situations where ensuring that perpetrators are held accountable for crimes have to be compromised for the sake of securing peace through amnesty, bargains, or political immunity. This is because in some African countries perpetrators often enter into negotiations and demand immunity as a guarantee before they sign peace agreements. In such instances, it is often asked how peace can be achieved without compromising efforts to ensure accountability for past wrongs. This peace and justice dichotomy has been a subject of intense controversy among scholars, practitioners, and policymakers due to the moral, legal, and political choices involved in ending conflict.

In contrast, proponents of the restorative justice models emphasize the need to repair, rebuild, and restore damaged social relationships and socioeconomic welfare as an essential element in fostering sustainable peace (Adebayo et al. 2014; Brewoo and Abdallah 2014; Danso and Osei-Tutu 2014). The examples of the truth and reconciliation commissions in South Africa, Liberia, and Ghana, the Gacaca courts in Rwanda and Palava Hut in Liberia, and the granting of blanket amnesty in Mozambique in 1992 after the civil war provide compelling insights into how such restorative models have been deployed to pursue justice (Murithi 2002; Doe 2009; Mutisi 2009). Nonetheless, some have questioned the value and efficacy of truth commissions, especially whether they are able to meet the expectations of their presumed beneficiaries. In more specific terms, skeptics argue that reconciliation processes do not necessarily lead to forgiveness and healing. Other critics have raised questions about the retributive elements of some of the customary justice systems in which perpetrators are punished after being found guilty. The widespread use of amnesty has also received some criticism in the literature as it is said to deny victims the right to redress for past wrongs (Davidheiser and Nyiayaana 2011; Ushie 2013).

Despite the vigorous debates, there is a strong and growing consensus that retributive justice models should operate alongside or compliment restorative justice models to address different levels and dimensions of truth and justice, and to spread a broader awareness and understanding of the nature of past atrocities. Given their limits, a holistic approach that combines both retributive and restorative justice models is seen as the best approach to transitional justice. However, limited literature exists on the impact of these transitional justice models on the societies in question, although anecdotal evidence suggests a modicum of stability in some contexts, such as South Africa, Liberia, and Sierra Leone. Moreover, applying a "gender lens" to these two models is becoming increasingly common in the literature due to the impact of war on women in conflict situations.

Connections and disconnections to global discourses on peacebuilding

African scholars have not offered anything qualitatively distinct from their counterparts elsewhere in the world. The various ideas and perspectives of African scholars discussed in the preceding sections are not too different from the main themes in the global peacebuilding debate. Arguably, part of the reason is that most of the African scholars are trained in Western institutions and depend on Western literature for their research. A significant level of convergence thus exists between African research and global discourses on many a peacebuilding issues.

First, there is some level of convergence about certain key principles of building peace, such as national and local ownership. National ownership is seen as a vital element of the success and sustainability of peacebuilding activities as it ensures that processes are nationally driven. This is an issue that is often emphasized in the mainstream literature on peacebuilding and more significantly, in the policy documents of national governments, African regional organizations, the UN, and external actors. The report of the 2015 review of the UN's peacebuilding architecture, for instance, argued for "inclusive national ownership" in peacebuilding, which implies participation by national governments, community groups, women representatives, youth, labor organizations, political parties, the private sector, CSOs, and minority groups (United Nations 2015a).

Second, another connection between African peacebuilding research and global discourses is the strong emphasis placed on gender, women, peace, and security. Over the past decade, African researchers, policymakers, multilateral organizations, and external partners have emphasized the goal of gender equality and women's empowerment in Africa to address key challenges such as poverty, inequality, violence against women and girls, and the underrepresentation of women in politics, leadership, and management-level positions in the public and private sectors. These discussions have led to gender being treated as a cross-cutting issue to be mainstreamed in all activities. Illustrating of the importance of this issue, the fifth Sustainable Development Goal (SDG) specifically targets achieving gender equality and empowering all women and girls by 2030.

Third, criticism of current liberal peacebuilding approaches that emphasize political liberalization (democracy building) and economic liberalization (market economics) as the surest foundations for peace is another similarity between African peacebuilding research and global discourses (Murithi 2012). In recent times, there has been increasing skepticism among African and Western scholars of whether the pursuit of liberal peace contributes to sustainable peace in post-conflict societies due to the experiences of African countries such as Liberia, Burundi, and Cote d'Ivoire, which relapsed into conflict. The global practices of peacebuilding are gradually shifting away from market-oriented policies, top-down reforms, and liberal democratization to focus more on locally based approaches that lead to context specific programs and better outcomes. For example, in the area of SSR the notion of "hybridity" which stresses the relationships between the various formal and informal systems, actors, and beneficiaries involved in the security sector is increasingly gaining currency among donor partners and international organizations. With respect to transitional justice, most international actors are placing greater emphasis on African cultures and customary practices, although there appears to be some difficulty in determining how these can be fused into the institutional norms, designs, structures, and process that drive the formal transitional justice systems (Bagayoko, Hutchful and Luckham 2016; Aning and Aubyn 2018).

Strengths of the African peacebuilding literature

While acknowledging that research informs policy incrementally and over time, it can be argued that African peacebuilding scholarship has contributed to the formulation and rethinking of national, regional, and global policies and practices of peacebuilding. African academics, policymakers, and practitioners, for example, provided substantial input into all the three major peace and security reviews conducted in 2015 by the UN on peace operations, peacebuilding, and UNSCR 1325 (United Nations 2015b).[7] This influence is reflected in some of the recommendations put forward by the reports. For example, one of the important issues advocated by a number of African researchers and institutions like the AU is local ownership of peacebuilding processes (African Union 2015). Thankfully, among the overarching recommendations in the report of the advisory group of experts for the 2015 review of the UN peacebuilding architecture was the need to foster "inclusive national ownership" through a common understanding between the state and all sectors of civil society to ensure that peace becomes sustainable.[8] Today, many of the UN, regional, and other external interventions in Africa put major emphasis on local ownership, although this is yet to be fully put into practice. It is therefore appropriate to give credit to African scholars and practitioners who have over the years advocated in their writings and policy engagement for local content and ownership of peacebuilding processes.

Additionally, the literature produced by African academics, practitioners, and institutions on different aspects of peacebuilding, such as democratic elections, good governance, SSR, market reforms, transitional justice, and reconciliation, has provided a valuable resource and reservoir of knowledge for policymakers and practitioners on key issues and lessons learned about peacebuilding in Africa. Another significant strength is that the analysis provided in the literature is often based on firsthand experience and knowledge, and local realities due to their geographical proximity. In some cases, the researchers even live in the communities where the research is being carried out. This gives them a comparative advantage in terms of access to the empirical data. Nonetheless, it is important to note that despite the fact that peace is largely built at the grassroots level, some of the African publications are produced by scholars with little or no touch with the communities where peace is built. Hence, their conclusions and recommendations are often not in tune with the real needs of the African people.

Weaknesses of the African peacebuilding literature

A review of the literature reveals some apparent limitations. To begin with, most of the critical and analytical perspectives that inform global peacebuilding policies and practices have largely been dominated by Western academics and institutions. Thus, there is a paucity of conceptual and theoretical research on African approaches to peacebuilding that can inform regional and international peacebuilding agendas. Consequently, African scholars engage in intellectual debates originating from the dominant peacebuilding discourses from the West. Despite criticizing Western peacebuilding policy prescriptions, scholarship, and practices, African researchers have not been able to clearly articulate strong countervailing normative frameworks. This lacuna contributes to the reinforcement of Western prejudices toward contemporary peacebuilding scholarship in Africa (Murithi 2012).

Another key weakness is that, with a few exceptions, the literature does not benefit from the most recent data on peacebuilding at the policy level due to accessibility challenges. Many are unable to engage multilateral institutions where global peacebuilding policies are debated and developed, such as the UN headquarters in New York where the UN Peacebuilding Commission is located. Moreover, in terms of research focus, the regional impacts of national peacebuilding efforts have not gained much traction in the existing literature. With increasing regional security complexities, a comprehensive analysis of how regional conflict dynamics affect the consolidation of peace at the national level is not only relevant but crucial to understanding how transnational issues undermine the sustainability of peace in post-conflict environments. A typical example is Mali, where the achievement of lasting peace in the country also depends on the conflict dynamics within the Sahel region. Issues such as radicalization, violent extremism, terrorism, climate change, and conflict have an especially transnational character, which suggests that an intervention exclusively focused on the national level may have limited impact (Aning, Abdellah and Aubyn 2016).Another area that has not received adequate research attention is the role played by CSOs in shaping peacebuilding policies and processes in Africa. Although CSOs have been instrumental in the implementation of peacebuilding projects in Africa, there is a dearth of literature that focuses on how they have shaped peacebuilding approaches on the continent and engaged with broader global policy processes. The role of CSOs needs to be fully unpacked to unlock their potential for building and sustaining peace in Africa. More crucially, African academics, as argued by Olonisakin, have rarely placed the issue of leadership at the core of their analyses of peace and stability. Leadership factors and dynamics in Africa can either lead to conflict or create foundations for peace on the continent. It is also instructive to note that evidence-based research that clearly articulates the contribution of African scholarship to global policy debates on peacebuilding is also lacking. Peacebuilding research is also overwhelmingly focused on the security and political dimensions with little focus on the economic dimension and recovery process (Olonisakin 2012).

Conclusion

There has been a remarkable expansion in African scholarly writing on peacebuilding since 1992 when the term was popularized in Boutros-Ghali's Agenda for Peace report. African scholars from diverse academic disciplines have written on several aspects of peacebuilding from various perspectives and contexts using different methodological approaches. This chapter highlights African perspectives on some of the key trends, themes, and debates within the peacebuilding literature over the past decade and how they relate to and deviate from global discourses and processes. This chapter also identified some of the strengths and weaknesses of African peacebuilding literature as well as its contributions to global peacebuilding literature.

Although diverse perspectives are evident in the foregoing analysis of African peacebuilding literature, the dominant theme is the suitability and sustainability of the liberal governance framework within the African context. The liberal peace approach is widely criticized by African scholars for its state-centric, top-down approach to dealing with the needs of post-conflict countries. For some scholars, the top-down approach obscures the important contributions that African traditional systems can make toward advancing peace. They argue that the local context and informal actors and initiatives need to be prioritized in the peacebuilding process. Related to this is the issue of local ownership, another recurrent theme in the literature, based on the principle that peacebuilding processes should always be spearheaded and owned by locals in order to be sustainable. However, although local ownership is a key principle in the conceptualization of peacebuilding, many of the African researchers surveyed argue that it is yet to be reflected in practice. Importantly, although African peacebuilding research has contributed to the formulation and review of national, regional, and global policies and practices on peacebuilding, it also has some limitations. Addressing the identified limitations is vital to enhancing peacebuilding research in Africa.

Notes

1 This chapter was originally published as a working paper by the African Peacebuilding Network of the Social Science Research Council at https://www.ssrc.org/publications/view/an-overview-of-recent-trends-in-african-scholarly-writing-on-peacebuilding/

2 United Nations Peacebuilding Commission, "Countries Currently on the PBC Agenda." Accessed February 20, 2018, http://www.un.org/en/peacebuilding/.

3 United Nations, *An Agenda for Peace: Preventive Diplomacy, Peacemaking and Peacekeeping* (New York: UNSC, 1992).

4 "Preventive diplomacy" is defined as action to prevent disputes from arising between parties, to prevent existing disputes from escalating into conflicts, and to limit the spread of the latter when they occur; "peacemaking" is defined as action to bring hostile parties to agreement, essentially through such peaceful means as those foreseen in Chapter VI of the Charter of the United Nations; "peacekeeping" is the deployment of a UN's presence in the field, hitherto with the consent of all the parties concerned, normally involving UN military and/or police personnel and frequently civilians as well; "peacebuilding" is defined as action to identify and support structures, which will tend to strengthen and solidify peace in order to avoid a relapse into conflict (de Coning 2010, 12).

5 For more information, see The African Union (AU) Policy Framework on Transitional Justice in Africa (ATJF).

6 International Criminal Court, "Case Information Sheet: The Prosecutor v. William Samuel Ruto and Joshua Arap Sang," ICC-01/09-01/11, April 2016.

7 United Nations, *Uniting Our Strengths for Peace: Politics, Partnership and People–Report of the High-level Independent Panel on United Nations Peace Operations* (New York: United Nations, 2015); United Nations, *The Challenge of Sustaining Peace–Report of the Advisory Group of Experts for the 2015 Review of the United Nations Peacebuilding Architecture* (New York: United Nations, 2015); United Nations, *The Global Study on UNSCR 1325* (New York: United Nations, 2015).

8 African Union, *Common African Position on the UN Review of Peace Operations* (Addis Ababa: AU, 2015).

References

Aboagye, Festus. (1999) *ECOMOG: A Sub-Regional Experience in Conflict Resolution, Management and Peacekeeping in Liberia.* Accra: Sedco Publishing.

Aboagye, Festus. (2004) "The African Union Mission in Burundi: Lessons Learned from the First African Union Peacekeeping Operation Peacekeeping." *Conflict Trends,* Issue no. 2. pp. 9–15.

Adebajo, Adekeye. (2002) *Building Peace in West Africa: Liberia, Sierra Leone and Guinea-Bissau.* Boulder, CO: Lynne Rienner.

Adebajo, Adekeye. (2004) "From Congo to Congo: United Nations Peacekeeping in Africa after the Cold War." In Ian Taylor and Paul Williams (eds.), *Africa in International Politics, External Involvement on the Continent.* London: Routledge, pp. 195–212.

Adebajo, Adekeye and Ismail Rashid (eds.) (2004) *West Africa's Security Challenges: Building Peace in a Troubled Region*. London: International Peace Academy.

Adebayo, Akanmu G., Brandon D. Lundy, Jesse J. Benjamin and Joseph Kingsley Adjei (eds.) (2014) *Indigenous Conflict Management Strategies in West Africa: Beyond Right and Wrong*. Lanham, MD: Lexington Books.

Adedeji, Ebo. (2007a) "Liberia Case Study: Outsourcing SSR to Foreign Companies." In Laurie Nathan (ed.) *No Ownership, No Commitment: A Guide to Local Ownership of Security Sector Reform*. Birmingham: University of Birmingham, pp. 78–93.

Adedeji, Ebo. (2007b) "The Role of SSR in Sustainable Development: Donor Policy Trends and Challenges." *Conflict, Security & Development* 7, no. 1, pp. 27–60.

Adedeji, Ebo. (2007c) "Toward Sustainable Peace and Security in Africa: The Challenges of Security Sector Reform in Post-conflict Sierra Leone." In Festus Aboagye (ed.), *Complex Emergencies in the 21st Century: Challenges of New Africa's Strategic Peace and Security Policy Issues*. Pretoria: Institute for Security Studies, pp. 49–72.

Adedeji, Ebo. (2008) "Towards a Common ECOWAS Agenda on Security SectorGeneva Centre for the Democratic Control of Armed Forces (DCAF) Policy Paper no. 23

Adetula, Victor, Redie Bereketeab, and Olugbemi Jaiyebo. (2016) *Regional Economic Communities and Peacebuilding in Africa: The Experiences of ECOWAS and IGAD*. Uppsala: The Nordic Africa Institute.

African Union. (2006). "The AU Policy on Post Conflict Reconstruction and Development." Addis Ababa: African Union.

African Union Panel of the Wise. (2013). "Research Report: Peace, Justice, and Reconciliation in Africa Opportunities and Challenges in the Fight Against Impunity." New York: International Peace Institute.

African Union. (2013) "Peace, Justice, and Reconciliation in Africa Opportunities and Challenges in the Fight against Impunity." The African Union Series.

African Union. (2015) *Common African Position on the UN Review of Peace Operations*. Addis Ababa: African Union.

Albert, Isaac O. (2008) "Understanding Peace in Africa." In David Francis (ed.), *Peace and Conflict in Africa*. London and New York: Zed Books, pp. 31–45.

Albert, Isaac O. (2017) "Teaching Peacebuilding in African Universities: Bridging the Gap between the Field and the Classroom." SSRC African Peacebuilding Network Working Paper No. 13.

Albert, Isaac O. and Oluremi Albert. (2008) "Strategies for Peace Education Integration in ECOWAS Member States' Tertiary School Curricula." *Conflict Trends* Issue no. 3, pp. 11–17.

Aning, Kwesi. (1997) "The International Dimensions of Internal Conflict: The Case of Liberia and West Africa." CDR Working Paper 97/4.

Aning, Kwesi and Festus Aubyn. (2018) "Challenging Conventional Understanding of Statehood based on West African Realities." In Kwesi Aning, Anne M. Brown, Volker Boege and Charles (eds.), *Exploring Peace Formations: Justice and Security in Post-conflict States*, T. Hunt. London: Routledge, pp. 22–26.

Aning, Kwesi, Mustapha Abdellah and Festus Aubyn. (2016) "Responding to Boko Haram: Interrogating the Effectiveness of State and Regional Intervention Approaches." In James J. Hentz and Hussein Solomon (eds.), *Understanding Boko Haram: Terrorism and Insurgency in Africa*. London: Taylor and Francis, pp. 237–254.

Aning, Kwesi and Thomas Jaye. (2011) "Liberia: A Briefing Paper on the TRC Report." KAIPTC Occasional Paper 33.

Aning, Kwesi and Ernest Lartey. (2010) "Establishing the Future State of the Peacebuilding Commission Perspectives on Africa." Working Papers on the Future of the UN Peacebuilding Architecture. No. 1.

Aning, Kwesi and Naila Salihu. (2012) "Regional Approaches to Statebuilding II: The African Union and ECOWAS." In Mats Berdal and Dominik Zaum (eds.), *The Political Economy of Post-conflict Statebuilding: Power after Peace.* New York City: Routledge, pp. 256–282.

Anyidoho, Henry Kwami. (1997) *Guns over Kigali: The Rwandese Civil War-1994.* Accra: Woeli Publishing Services.

Bagayoko, Niagale, Eboe Hutchful and Robin Luckham. (2016) "Hybrid Security Governance in Africa: Rethinking the Foundations of Security, Justice and Legitimate Public Authority." *Conflict, Security & Development* 16, pp. 1–32.

Bah, Abu Bakarr. (2017) "People-Centered Liberalism: An Alternative Approach to International State-Building in Sierra Leone and Liberia." *Critical Sociology* 43, Issue 7–8. pp. 1–19.

Bah, Sarjoh A. (2010) *Dilemmas of Regional Peacemaking: The Dynamics of the AU's Response to Darfur.* New York: Center on International Cooperation, New York University.

Brewoo, Serwaa and Mustapha Abdallah. (2014) "Exploring Indigenous Mechanisms for Peacemaking in West Africa." In Akanmu G. Adebayo, Brandon D. Lundy, Jesse J. Benjamin and Joseph Kingsley Adjei (eds.), *Indigenous Conflict Management Strategies in West Africa: Beyond Right and Wrong.* Lanham, MD: Lexington Books. pp. 38–56.

Bryden, Alan, Boubacar N'Diaye, and Funmi Olonisakin (eds.) (2008) *Challenges of Security Sector Governance in West Africa.* Zurich: LIT Verlag.

Bussy, Josaphat Musamba, and Carol Jean Gallo. (2016) "The Great Lakes Region of Africa: Local Perspectives on Liberal Peacebuilding from the Democratic Republic of Congo." In Oliver P. Richmond, Sandra Pogodda and Jasmin Ramovic (eds.), *The Palgrave Handbook of Disciplinary and Regional Approaches to Peace.* London: Palgrave Macmillan. pp. 312–324.

Curtis, Devon, and Gwinyayi A. Dzinesa. (2013) *Peacebuilding, Power and Politics in Africa.* Johannesburg: University of the Witwatersrand Press.

Danso, Kwaku. (2016) "Mending Broken Relations after Civil War: The 'Palava Hut' and the Prospects for Lasting Peace in Liberia." KAIPTC Policy Brief 2.

Danso, Sarah, and Joana Osei-Tutu. (2014) "Homegrown Crises, Homegrown Solutions? The Efficacy of Indigenous Conflict Resolution/Management Approaches in Ghana." In Akanmu G. Adebayo, Brandon D. Lundy, Jesse J. Benjamin and Joseph Kingsley Adjei (eds.), *Indigenous Conflict Management Strategies in West Africa: Beyond Right and Wrong.* Lanham, MD: Lexington Books, pp. 148–165.

Davidheiser, Mark and Kialee Nyiayaana. (2011) "Demobilization or Remobilization? The Amnesty Program and the Search for Peace in the Niger Delta." *African Security* 4, no. 1, pp. 44–64.

de Coning, Cedric (2010) "Clarity, Coherence and Context: Three Priorities for Sustainable *Peacebuilding.*" *Working Papers on the Future of the UN Peacebuilding Architecture* no. 3, pp. 1–28.

de Coning, Cedric. (2016) "From Peacebuilding to Sustaining Peace: Implications of Complexity for Resilience and Sustainability." *Resilience: International Policies, Practices and Discourses*, 4, Issue no. 3, pp. 166–181.

Doe, Samuel Gbaydee. (2009) "Indigenizing Post Conflict State Reconstruction in Africa: A Conceptual Framework." *Africa Peace and Conflict Journal*2, no. 1, pp. 1–17.
Eghosa, Osaghae E. (2001) "The Role and Function of Research in Divided Societies: The Case of Africa." In Marie Smyth and Gillian Robinson (eds.), *Researching Violently Divided Societies: Ethical and Methodological Issues.* London: Pluto Press, pp. 12–33.
Ero, Comfort. (1995) "ECOWAS and the Sub Regional Peacekeeping in Liberia." *Journal of Humanitarian Assistance*, September 25, pp. 1–19
Fombad, Charles Manga. (2008) "Transitional Justice in Africa: The Experience with Truth Commissions." http://www.nyulawglobal.org/globalex/Africa_Truth_Commissions.html.
Francis, David J. (2000) "ECOWAS Cease-fire Monitoring Group (ECOMOG): A New Security Agenda in World Politics." In Bakut Simon and Sagarika Dutt (eds.), *Africa towards the Millennium: An Agenda for Mature Development.* London: Palgrave, pp. 177–202.
Francis, David J. (2007) "Peace and Conflict Studies: An African Overview of Basic Concepts." In Shedrack Gaya Best (ed.), *Introduction to Peace and Conflict Studies in West Africa.* Ibadan: Spectrum Books Limited, pp. 14–27.
Francis, David J. (2009) "Peacekeeping in a Bad Neighbourhood: The Economic Community of West African States (ECOWAS) in Peace and Security in West Africa." *African Journal on Conflict Resolution* 9, no. 3, pp. 87–116.
Galtung, Johan. (1976) "Three Approaches to Peace: Peacekeeping, Peacemaking, and Peacebuilding." In Johan Galtung (ed.), *Peace, War and Defense: Essays in Peace Research.* Copenhagen: Christian Ejlers, pp. 282–304.
Gbla, Osman. (2007) "Security Sector Reform in Sierra Leone." In Len le Roux and Yemane Kidane (eds.), *Challenges to Security Sector Reform in the Horn of Africa.* Pretoria: Institute for Security Studies, pp. 13–36.
James-Allen, Paul, Aaron Weah, and Lizzie Goodfriend. (2010) "Beyond the Truth and Reconciliation Commission: Transitional Justice Options in Liberia." International Centre for Transitional Justice.
Jaye, Thomas. (2003) *Issues of Sovereignty, Strategy and Security in the Economic Community of West African States (ECOWAS) Intervention in the Liberian Civil War.* New York: Edwin Mellen Press.
Jaye, Thomas. (2008) *Liberia's Security Sector Legislation.* Geneva: DCAF.
Jaye, Thomas. (2009) "Transitional Justice and DDR: The Case of Liberia." Research Unit: International Center for Transitional Justice Working Paper.
Jaye, Thomas, Dauda Garuba, and Stella Amadi (eds.) (2011) *ECOWAS & the Dynamics of Conflict and Peacebuilding.* Dakar: CODESRIA.
Kagwanja, Peter. (2017) "Kenya's 2017 Electoral Impasse a Severe Test to Liberal Peace-Building." *Daily Nation* (Kenya), October 1.
Karbo, Tony. (2012) "Localising Peacebuilding in Sierra Leone: What Does It Mean?" Conference Paper, Issue 3.
Kariuki, Nahashon. (2014) *Transitional Justice in Kenya: A Historical Perspective and a Synopsis of a Troubled Truth Justice and Reconciliation Commission.* Nairobi: NPI-Africa and GPPAC.
Lederach, John Paul. (1997) *Building Peace: Sustainable Reconciliation in Divided Societies.* Washington, DC: U.S. Institute of Peace Press.
Lemay-Hebert, Nicolas. (2013) "Review Essay: Critical Debates on Liberal Peacebuilding." *Civil Wars* 15, no. 2, pp. 242–252.
Malan, Mark. (2008) *Security Sector Reform in Liberia: Mixed Results from Humble Beginnings.* Carlisle, PA: Strategic Studies Institute, US Army War College.

Mamdani, Mahmood. (2016) "Who's to Blame in South Sudan?" *Boston Review*. Accessed on September 17, 2016. http://bostonreview.net/world/mahmood-mamdani-south-sudan-failed-transition.

McCandless, Erin, Abdul Karim Bangura, Mary E. King, and Ebrima Sall. (2007) *Peace Research for Africa: Critical Essays on Methodology*. Addis Ababa: UPEACE Africa Programme.

McCandless, Erin, and Tony Karbo. (2011) *Peace, Conflict, and Development in Africa: A Reader*. Geneva: University for Peace.

Murithi, Tim. (2002) "Rebuilding Social Trust in Northern Uganda." *Peace Review* 14, no. 3, pp. 291–295.

Murithi, Tim. (2006) "African Approaches to Building Peace and Social Solidarity." *African Journal of Conflict Resolution* Vol. 6, no. 2, pp. 9–34.

Murithi, Tim. (2008) "The African Union's Evolving Role in Peace Operations: The African Union Mission in Burundi, the African Union Mission in Sudan and the African Union Mission in Somalia." *African Security Review* 17, no. 1, pp. 70–82.

Murithi, Tim. (2012) "The State of Research on Peacebuilding in Africa." African Peacebuilding Network (APN) Draft Scoping Study.

Mutisi, Martha. (2009) "Gacaca Courts in Rwanda: An Endogenous Approach to Post-Conflict Justice and Reconciliation." *Africa Peace and Conflict Journal* 2, no. 1, pp. 17–26.

Mutisi, Martha. (2012) "Interrogating Traditional Youth Theory: Youth Peacebuilding and Engagement in Post-conflict Liberia." In Grace Maina (ed.), *Africa Dialogue Opportunity or Threat: The Engagement of Youth in African Societies*. Durban: ACCORD, pp. 87–120.

Nwolise, Osisioma B. C. (1992) "The Internationalization of the Liberian Crisis and Its Effects on West Africa." In Margaret Vogt (ed.), *The Liberian Crisis and ECOMOG: A Bold Attempt at Regional Peacekeeping*. Lagos: Gambuno Press, pp. 63–78.

Nyokabi, Kamau, Ekaette Ikpe, and Abiodun Alao. (2014) "Building the State without Peace or Making Peace without the State: A Baseline Study on the Paradox of Statebuilding and Peacebuilding in Sierra Leone." Paper presented at the Project's Baseline Study Review Workshop Convened on June 26–27, 2014, Nairobi.

Obi, Cyril. (2009) "Economic Community of West African States on the Ground: Comparing Peacekeeping in Liberia, Sierra Leone, Guinea Bissau, and Côte D'Ivoire." *African Security* 2, nos. 2–3, pp. 119–135.

Okyere, Frank. (2015) "Statebuilding, Local Governance, and Organized Crime in Mali." In Andrea Ó Súilleabháin (ed.), *Leveraging Local Knowledge for Peacebuilding and Statebuilding in Africa*. New York: International Peace Institute, pp. 24–31.

Olonisakin, Funmi, Alagaw Ababu, and Alfred Muteru. (2017) "Thinking about Peace and the Role of State Building and Political Settlement in the Pursuit of Sustainable Peace in Africa." IDRC Working Paper. Accessed on October 15, 2017. https://idl-bnc-idrc.dspacedirect.org/bitstream/handle/10625/56335/IDL-56335.pdf?sequence=2.

Olonisakin, Funmi, and Alfred Muteru. (2014) "Reframing Narratives of Peacebuilding and Statebuilding in Africa." ALC Working Paper, February.

Olonisakin, Funmi. (2012) "Leadership and Peacebuilding in Africa." ALC Working Paper, March.

Olonisakin, Funmi. (2008) *Peacekeeping in Sierra Leone: The Story of UNAMSIL Histories of UN Peace Operations*. Boulder, CO: Lynne Rienner Publishers.

Olowu, Dele, and Paulos Chanie (eds.) (2016) *State Fragility and State Building in Africa: Cases from Eastern and Southern Africa.* New York: Springer.

Omeje, Kenneth C. (2015a) "Promoting Peace and Conflict-Sensitive Higher Education in Sub-Saharan Africa." *Africa Conflict and Peacebuilding Review* 5, no. 2, pp. 33–56.

Omeje, Kenneth C. (2015b) "Strengthening Peace Research and Peace Education in African Universities." *African Sociological Review* 19, no. 1, pp. 16–33.

Rashid, Ismail. (2017) "Writing on African Peacebuilding: Reflection on Personal Experiences." Kujenga Amani. Accessed on October 17, 2017. https://kujenga-amani.ssrc.org/2017/01/09/writing-on-african-peacebuilding-reflection-on-personal-experiences/.

Salih, Mohamed A. (2017) "A Critique of the Political Economy of the Liberal Peace: Elements of an African Experience." In Edward Newman, Roland Paris and Oliver P. Richmond (eds.), *New Perspectives on Liberal Peacebuilding.* Tokyo: United Nations University Office, pp. 133–158.

United Nations. (1992) *An Agenda for Peace: Preventive Diplomacy, Peacemaking and Peacekeeping.* New York: UNSC.

United Nations. (2000) *Report of the Panel on United Nation Peace Operations of 2000.* New York: UNSC.

United Nations. (2004) *Report of the High-level Panel on Threats, Challenges and Change.* New York: UNSC.

United Nations. (2015a) *The Challenge of Sustaining Peace – Report of the Advisory Group of Experts for the 2015 Review of the United Nations Peacebuilding Architecture.* New York: UNSC.

United Nations. (2015b) *Uniting Our Strengths for Peace: Politics, Partnership and People – Report of the High-level Independent Panel on United Nations Peace Operations.* New York: UNSC.

Ushie, Vanessa. (2013) "Nigeria's Amnesty Programme as a Peacebuilding Infrastructure: A Silver Bullet?" *Journal of Peacebuilding & Development* 8, no. 1, pp. 30–44.

Von Gienanth, Tobias, and Thomas Jaye. (2007) *Post-Conflict Peacebuilding in Liberia: Much Remains to Be Done.* Berlin: Center for International Peace Operations (ZIF) and Kofi Annan International Peacekeeping Training Centre (KAIPTC).

West African Network for Peacebuilding (WANEP). (2017) *Strides and Strains of Civil Society Organizations in West Africa: The WANEP Story.* Accra: WANEP.

Zambakari, Christopher. (2016) "Challenges of Liberal Peace and Statebuilding in Divided Societies." *Conflict Trends Issue* no. 4. https://www.accord.org.za/conflict-trends/challenges-liberal-peace-statebuilding-divided-societies/

Zondi, Siphamandla. (2017) "African Union Approaches to Peacebuilding Efforts at Shifting the Continent towards Decolonial Peace." *African Journal of Conflict Resolution* 17 no. 1.pp. 105–131.

2 When theory meets method

Feminist peace research in Africa and how to make the strange familiar and the familiar strange

Heidi Hudson

Peacebuilding practices are usually driven by the goal to make policy-related and practical recommendations regarding specific peacebuilding projects, and not to engage in systematic theory-building. Such operations are also mainly characterized by emergency relief, stabilization and development imperatives, and are therefore not strongly rooted in conflict resolution theory (Bercovitch and Jackson 2009, p. 179). Within this context, the prevailing peacebuilding literature reflects a binary between liberal peacebuilding work with a set of fairly coherent core assumptions (Paris 2004, 2010) and a rather scattered collection of critical peacebuilding research interventions (Autesserre 2014; Chandler 2009; Duffield 2001; Richmond 2011). The lack of a coherent body of critical theory therefore has consequences for the ability of critical peacebuilding research to identify and explain what the deeper causes of conflict are and how specific targeted peacebuilding strategies can be designed to address those causes (see Bercovitch and Jackson 2009, p. 179). Unlike liberal peacebuilding that boasts a rather consistent and enduring framework for political reform, critical peacebuilding alternatives have not risen to the challenge to offer sustained answers to the very questions they raise.

For this reason the focus of this chapter is on the intersection between theory and method and the consequences when they meet in the field – specifically in peacebuilding practices in Africa. Postcolonial approaches have been viewed as an alternative to overcome the inability of liberal approaches to challenge power differentials between the local and the international (Bhabha 1994; Said 1978). But postcolonialism has also lately come under fire from decolonial scholars (Grosfoguel 2007; Mignolo 2011; Ndlovu-Gatsheni 2014; Quijano 2007) who argue that it is difficult to disentangle postcolonial theory from the coloniality embedded in its analytical tools. According to these scholars, a decolonizing theory cannot stem from the West – it has to be one that originates from its margins or borders, where the oppression is felt (Hudson 2016, p. 198). Meera Sabaratnam (2017) advocates in this regard a decolonial approach that integrates the voices of the targets of intervention by paying attention to their historical presence, political consciousness and material conditions. The implication for a theory of peacebuilding is therefore that attempts to overcome the international-local binary of liberal peacebuilding through the notion that "liberal-local hybridity" (Richmond 2010, pp. 14–38) is regarded as being unable to shift the terms of the engagement far enough, since such a postcolonial critique remains lodged within a Eurocentric frame of thinking. In response, postcolonial scholars defend hybridity (or the connectedness of the global and the local) not as an acceptance of the superiority of Western knowledge, but rather as an acknowledgement that the production of indigenous or local theory is often wrongly presented as pure and uncontaminated (Acharya and Buzan 2010, p. 227).

In view of a rather unproductive standoff between scholars of decoloniality and "the rest" (including critical theorists of postcoloniality), I turn in this contribution to feminist scholarship for the way in which it integrates theory and praxis. Feminist scholarship in general and feminist International Relations (IR) have a long history of reflection on many of the contemporary issues related to peacebuilding such as a bottom-up logic. My aim is to show how a multi-pronged feminist methodology informed by postcolonial and poststructuralist feminist theory can invigorate peacebuilding theorizing and practice in a way that does not entrench Eurocentric assumptions but challenges all kinds of epistemological and methodological binaries. I take inspiration from feminist research principles of critical reflexivity, grounded in interdisciplinarity, concrete (everyday) experience and the recognition that research is always and already political (e.g. Harding 1986). I pay specific attention to two feminist analytical tools, namely discourse analysis as a tool of poststructuralist feminism and intersectionality as a key feature of postcolonial feminism. I submit that since all discursive and non-discursive peacebuilding practices form part of an entangled global coloniality that is simultaneously radicalized, patriarchal, hetero-normative and capitalist, among others (Ndlovu-Gatsheni 2014, pp. 187–188), gender is an important but not always the primary focus in peacebuilding. The two critical-feminist methodological and epistemological "tools" help us to understand this fact by focusing on how these entanglements play out in specific conflict and post-conflict contexts, thereby subverting familiar peace and security knowledges (making them strange) while also "familiarizing the strange."[1] In so doing, these critical feminist theories and their respective methods show us a way forward to overcome the binary deadlock between the colonizer and the colonized. I contend that the combination of critical discourse analysis (CDA) and intersectional analysis guided by feminist ethics and reflexivity offers a valuable entry point into decolonized methodologies for peacebuilding research.

In the first part of this chapter, I briefly establish the historical evolution of core liberal peace principles that have provided normative and conceptual/theoretical coherence to liberal peacebuilding – arguing that these principles have promoted its longevity as well as its ability to adapt to criticism. In this section, I also draw attention to the failed attempts by critics to offer alternatives. I conclude that part of the reason why they have been unsuccessful is because of the neglect of a critical feminist perspective. The second part of the chapter therefore hones in on poststructuralist and postcolonial feminist approaches – outlining the limitations of liberal-feminist theory and methods, and making a case for blending the assumptions and methods of critical feminist approaches. In the final section, I illustrate the applicability of these theories and methods through examples of feminist research on peacebuilding that straddles war, postwar and everyday war in Africa, with particular focus on the construction of sexual violence in the Democratic Republic of Congo (DRC).

The secrets of the liberal peace success

Liberal mainstream approaches to peacebuilding are difficult to decenter for a number of reasons. First, policy-makers and practitioners have been able to maintain a core set of principles while also being quite pragmatic in their "responsiveness" to critique. Neoliberal governance has become more politically correct. Second, critics have failed to come up with alternatives that are decolonial at their core (Sabaratnam 2013, pp. 259–260). In this section, I explore these claims in more detail.

With regard to the first reason, it is worth noting that the meaning of concepts is always open to political and ideological manipulation, and can therefore change over time depending on the prevailing international system, the preoccupations of its architects (Mac Ginty and Williams 2009, pp. 126–127) as well as the pragmatic and strategic adjustments. Such was the case with the evolution of the meaning of the concept "peacebuilding." During the Cold War, traditional peacekeeping was the norm, but since the end of the Cold War and Boutros Boutros-Ghali's 1992 *An Agenda for Peace* a distinction was made between peacekeeping, peace enforcement and post-conflict peacebuilding. Gradually the definition of peacebuilding was broadened to include not only preventive diplomacy and humanitarian aid, but also different types of civilian assistance, military operations, development initiatives and post-conflict reconstruction (Paris and Sisk 2009, p. 5). First generation peacebuilding efforts (e.g. in Bosnia) in the early part of the 1990s exposed the inexperience of the international community with post-conflict stabilization. Shifts in mandates from a short-term focus on post-conflict elections to more emphasis on achieving the conditions for stability (as in Kosovo and Timor Leste) gradually developed to allow second generation peacebuilding scholarship to focus more on the role of natural resources, transitional justice and women/gender in post-conflict recovery (Paris and Sisk 2009, p. 7).

Liberal peacebuilding thus became a popular label in the first half of the 21st century (Paris 2004, p. 5) and has subsequently been guided by a rather coherent set of liberal values related to the rule of law, markets and democracy. They form the cornerstones of the uniform design and implementation of a particular kind of (universalized) peace. Such design is underpinned by the belief that "to have legitimacy, the state must be organized around liberal-democratic principles, and that because liberal democracies are respectful of their societies and peaceful toward their neighbors, they are the foundation of a stable international order" (Barnett 2006, p. 88).

Ideological and epistemological coherence has further been promoted by the use of simplistic conflations, such as between peacebuilding and democratization, such as in Iraq and Afghanistan. The treatment of peacebuilding and statebuilding as synonymous further succeeded to entrench the connection between a particular type of international peace undergirded by the liberal state and economic liberalization (Roberts 2008, p. 541). Attention to the power of these conflations and connections helps to explain the durability of liberal peace theory and practices. Thus, rather than viewing liberal peacebuilding "theory" as being hampered by its inability to engage with context beyond their own, or its "methodological inability to move beyond its universal prescriptions derived from a narrow Western experience" (Richmond 2011, p. 3), I argue that it is the presence of universalized prescriptions that facilitated its longevity.

Within the context of such a "system" of interconnected values, practical adjustments are made in response to criticism as well as to the realities on the ground. It therefore remains an open question whether the local-liberal hybrids which are beginning to take shape actually testify to the existence of postliberal elements within the dominant framework or whether they are examples of political correctness (see for instance the literature on hybrid political orders; Boege et al. 2008) and the introduction of gender-relational approaches to peacebuilding (Myrttinen, Naujoks and El-Bushra 2014).

The second reason for the enduring success of liberal peacebuilding as a dominant discourse and practice relates to the uneven development of a so-called "postliberal peace alternative" (Richmond 2011). Normatively the critical peacebuilding project is loosely kept together by its critique of the asymmetrical power relations between external and internal stakeholders (Heathershaw 2008, p. 604). Critical approaches to peacebuilding are informed by principles of social justice and emancipation, and local ownership founded on principles of communitarianism. The research agenda includes attention to the dynamics of postcolonial hybrid forms of peace. These include the (everyday) ways in which peace is constructed and negotiated (co-constituted) between locals and the international community, focusing on forms of local resistance as manifestations of an "infrapolitics of peacebuilding" (Jennings 2014; Richmond 2011, pp. 14, 16). However, the key question is whether these critical positions offer stand-alone alternatives to liberal peacebuilding or whether they are also responses that are drawing on liberal principles, but which lack an implementable plan to protect vulnerable people from lawlessness and predation (Paris 2010, p. 338).

One should therefore not underestimate the political nature of all grand projects, and remain vigilant to the fact that peace governance draws on a cobweb of interconnected power imbalances within the global political economy and security complexes. Although concepts such as "everyday peacebuilding" (Richmond 2009) and empathy/care are useful, they are not new, and peace scholars would do well to look more closely at how feminist scholarship has long theorized the ethics of care and the everyday of war and peace as simultaneously personal, political and international (Enloe 1990; Robinson 2011; Sylvester 2011). In the next section, I explain how a critical gender lens facilitates such integration of theory and method.

The indivisibility of feminist theory and method: lessons for critical peacebuilding research

Drawing on the long history of dialectical relationship between theory and practice in feminist scholarship as well as the foundations of standpoint feminism, I argue that such indivisibility offers useful pointers for conducting peacebuilding research from a critical perspective. This merging of theory and practice in feminist security thinking was influenced by, on the one hand, lessons learnt from the intersection between women's involvement in peace movements (e.g. the International Congress of Women at The Hague) and gender analysis (Tickner 1994, p. 44). On the other hand, standpoint feminism with its emphasis on situated knowledge, power inequalities and ethics of reflexivity also informed the recognition that a feminist intellectual conceptualization of security is grounded in the empirical authenticity of Black women's experiences within their everyday social worlds (Hill Collins 1997; Reynolds 2002).

Critical gender and feminist approaches to peacebuilding form part of a post-liberal pushback. Since the 1980s feminist IR scholars have challenged the focus on state and military (militarized) security and have sought to reconceptualize peace and security through a gender lens (e.g. Enloe 1990, 2007; Reardon 1993; Tickner 1992). More recently, Feminist Security Studies, a relatively new IR subfield, has begun to explore the ways in which political economy and human security approaches incorporate gender issues and consider women as subjects or agents of security. With the emphasis on the experiences of ordinary people, this field is increasingly concentrating on "the gray zones between war and peace, conflict and post-conflict" (Sylvester 2013a, p. 618). These scholars have generally been critical of the epistemically violent ways in which the liberal peace project employs liberal-feminist discourses of gender equality and mainstreaming to help enforce its norms and shape its practices on the ground (Pratt and Richter-Devroe 2011; Shepherd 2008, 2011).

Critical gender research on war and peace follows the logic of challenging binaries (power relations) and othering practices in specific conflict and peacebuilding contexts. Thus when feminist research exposes how masculinity is privileged over femininity or the public is valued over the private, we see how gender acts as an apparatus or political structure that determines "how actors order the world and how actors are ordered by it" (Gentry 2016, p. 20). We also then begin to understand how these historically and culturally constructed gender binaries operate to relegate women to being peaceful nurturers and men to being violent actors (Richter-Montpetit 2016, p. 95).

Although gender does not stand alone but rather functions in relation to other structures (as will be discussed later), if gender is the main category of analysis, then what one sees of gender is directly related to the type of gender lens used, be that liberal, radical/cultural or poststructuralist, for instance. In each of these theories a specific gender strategy is followed, namely of sameness, difference or diversity respectively. The first theory, liberal feminism, conceptualizes gender equality on the basis of a strategy of sameness. Consequently, it rejects the female stereotype and argues that women should aim to become more like men (Wibben 2016, p. 5). The second theory/approach accepts the stereotypes of women being more peaceful than men and pushes in a pacifist way for ending militarization to achieve a peace grounded in feminine values. The approach draws on cultural/radical feminism that uses gender as a strategy to affirm difference from the male norm, and challenges the assumption that women should imitate this norm. The third position is based on poststructuralist feminism, which views feminist analytical categories as unstable and fluid. This approach therefore rejects both male and female stereotypes and foregrounds strategies of diversity or multiplicity (of women's and men's stories and experiences of war and peace) to transform gender relations and structures altogether as part of a non-patriarchal and non-militarist global order (Wibben 2016, pp. 5–7).

In the same way that the core assumptions of the feminist theory determine the type of gender strategy or lens, so does the choice of lens influence the selection of research methods. Most feminist research on peace and war is characterized by historical or cross-cultural case studies with a strong ethnographic slant (Sylvester 2013b). Feminist research places a high premium on the notion of situated knowledge (Haraway 1988) that is partial, relational and subjective. It means that the knowledge produced by and through women's experiences of everyday peace and (in)security interacts with knowledge produced by others in that context. The addition of specific gender lenses adds nuance, allowing researchers to draw on a rich number of intellectual sources/theories to thicken their methodologies. For instance, following a poststructuralist-feminist approach that is wary of any essentialist assumptions may offer openings to uncover empirical evidence of how some women in the military conform to masculinist militarist ideals while at the same time challenging institutional gender stereotypes. Similarly, following a difference or standpoint approach (although running the risk of essentializing women's experiences) might provide insight into how a feminine peacefulness can be harnessed to resist war (Wibben 2016, p. 5).

If one takes all these theoretical feminist choices seriously, it has profound ethical implications for the kind of research questions posed as well as for the kinds of research practices one adopts. Fundamentally it means that no matter which approach is adopted critical feminist research is reflexive and always already concerned about "whose aims and interests the research is serving"

(Wibben 2016, p. 10). Annick Wibben reminds us that reflexive practices consists of two kinds of attentiveness to the politics of research, namely in relation to the limits of knowledge generation in a universal sense as well as with regard to the role of the researcher as a knowledge producer (Wibben 2016, p. 10). Recognizing that our values as researchers shape all aspects of the research process is a fundamental step in the process of unlearning patriarchal and imperialist research practices which privilege the authority of the researcher.

In the following two sections, I hone in on two feminist theories and their distinctive methodological tools (the discursive methods of poststructuralist feminism and the intersectional lens of postcolonial feminism). It does not however mean that these methodologies are used exclusively by these theories. In the final section, I illustrate how their integration in a practical case study from Africa could work toward supporting the arguments advanced previously.

Security language matters: poststructuralist feminism and the discursive turn

Both feminist poststructuralism and postcolonial feminism employ the methodological and epistemological tools of discourse analysis to deconstruct reality, and to show how language and reality are often performatively co-constituted (Gentry 2016, p. 24). This trend has also become popular in Feminist Security Studies research (Shepherd 2008) where the discursive construction of violence and security (through written, spoken or visual forms and texts) reveals hidden formulations of power in security relations between locals and international peacebuilders and/or among local stakeholders.

Discourse entails historically, socially and institutionally specific structures of statements, terms, categories and beliefs, together with all the practices and materialities that create meaning and identity (Foucault 1972, p. 80). It therefore constitutes more than just language or words. Yet, language is important because it is a vehicle for interpretation and representation of, for instance, mainstream peacebuilding or security discourse. CDA (Fairclough 2012; Milliken 1999) applied through a specific gender lens is therefore a very useful tool to domesticate the strange and defamiliarize or "queering" the familiar. CDA and critical-feminist approaches share similar normative orientations, namely the concern for "who is speaking, who is listening" (Gentry 2016, p. 35). Both CDA and poststructuralist feminism are concerned about how power is constructed and maintained through language, and how such language becomes a conduit for perpetuating injustice and inequality (Gentry 2016, p. 35).

The value of a feminist CDA tool is two-fold. First, it allows for understanding how certain discourses such as the liberal peace become dominant and naturalized. For example, a poststructuralist-feminist approach would therefore try to explain "how" gender is framed in liberal peacebuilding discourses – drawing our attention to the construction of women and girl children as helpless victims who need male protection. This type of theoretical lens (see previous section) and its concomitant methodology/instrument (discourse analysis) help to make the operations of discursive power visible by revealing how security discourse "positions people in their relations to [their gendered selves]… and to other [gendered] human beings within a particular symbolic [gender] order" (Hudson 2012, p. 444). It helps us to understand how and why the liberal peace discourse and the liberal-feminist gender equality discourse go hand in hand. Second, feminist CDA promotes better understanding of the consequences of this discursive power. The discourses and practices of peacebuilding actors such as the UN and key Western powers not only shape and reproduce gendered or other power relations, but also gradually produce representational truths about the strange (the "other") as well as the familiar ("us"). Such discursive power then determines the gendered path of post-conflict peace and reconstruction processes in very material ways. Through CDA we therefore begin to see how the continual reproduction of specific meanings of "security othering" solidifies the chain between discursive, epistemic violence and harmful material practices on the ground.

Although the correlation between exclusionary language in international discourses and local everyday violence is a key theme in poststructuralist/critical thinking about peace, the fluidity of language also points to a less deterministic analysis. In my analysis of the gender and women-focused language of UN Security Council Resolution (UNSCR) 1325 (2000) National Action Plans from Africa (Hudson 2017), I found that the four plans from Kenya, Liberia, Nigeria and Uganda – to varying degrees – reflect an ambivalent mix of predominantly liberal-feminist language interspersed with some examples of critical insight. For this reason, one cannot simply dismiss the liberal-feminist language as harmful, and would need to engage in a CDA that is sensitive to context and the political dynamics of implementation.

In this research I applied a critical-feminist discourse analysis approach to the four African National Action Plans. I pre-selected three concepts, namely gendered identity, security and violence. The methodology involved reading the texts against the grain, i.e. contrary to their ostensible logic to expose taken-for-granted usage of these concepts. Three representational themes emerged – the making of "women and children," "women civilizing war," and "making women responsible for preventing gender-based violence," informed by the literature on feminist critiques of UN texts. Second and third readings were conducted to avoid the risk of prescriptively foregrounding one perspective – to allow "other" interpretations. An ambivalent narrative emerged through this process – revealing not only what was missing, but also encouraging a reading of the liberal-feminist slant in a more open-minded, productive and constructive way. I therefore find in this research that the ambivalent nature of the messages sent out by these plans is a resource rather than an impediment. It underscores the fragmented nature of discourses, which in turn offers an opening for transcending the shackles of the soft consensus language of the UNSCR. Ambivalence creates new spaces for feminist engagement with policy and practice.

Gender does not stand alone: postcolonial feminism and intersectional identity constructions

Although Black (hooks 1984), Chicana (The Santa Cruz Feminists of Color Collective 2014) and Third World postcolonial feminisms (e.g. Chowdhry and Nair 2004; Mohanty 1988) also employ similar critical discourse-analytical methods, these theories are most recognized for their complex intersected understanding of, for instance, race, class, gender, sexuality, ethnicity and disability identity categories. In this conceptualization gender identity neither stands alone, nor is it always the primary category of analysis (Crenshaw 1991; hooks 1984), especially not in the global South. Thus, these feminisms challenge the assumption that women of the global South are homogeneously different, defined by their passivity and victimhood. An intersectional feminist approach underlines the fact that overlapping identity categories such as race and gender are not necessarily equal and that their relative importance is context-specific (Wibben 2016, p. 3). It means that women's gendered experiences of peacebuilding must be understood as "one of a series of [experiences] that shape subjectivities and thus the positionality of women" (Wibben 2016, p. 4).

Since agency is dependent on the "mix" of identities/subjectivities and the fact that women's and men's experiences of war or peace are not only the result of their gender, one should also view the causes of war and peace as complex and not only linked to gender inequalities. In this regard, Cori Wielenga's (2014) research through life narratives revealed that in Rwanda, there are many ways in which people experience their lived identities. Contrary to international perception, ethnicity is only one of many identities, but not necessarily primary. It follows that in a peacebuilding context, one should be critical of unproblematized gender roles of women as peaceful mothers and victims and men as protectors or aggressors as well as stereotypical assumptions that a female presence in the security sector will necessarily result in positive change. This is not only essentialist but also masks the victimhood of men and boys. So-called "gender-relational" approaches have recently taken root within peacebuilding discourse and practices, where the interplay between gender and other identity markers, such as age, social class, sexuality, disability, ethnic or religious background, marital status or urban/rural setting, is emphasized (Myrttinen, Naujoks and El-Bushra 2014).

These developments acknowledge the absence of sexual and gender minorities, including intersex, transgender and third gender persons, and masculinities from peacebuilding discourse, programs and policies, but are still couched in liberal-feminist language that policy-makers and practitioners can digest. Although these developments are considered positive, they do not as a rule reflect postcolonial or decolonial-feminist orientations that display a full consideration of historical and political contexts of struggle. These gender-relational approaches could very well be examples of liberal peacebuilding pragmatism which stand in contrast to deep (decolonial) engagement with gendered nationalisms, gendered militarisms and a violent postcolonial state as root causes of insecurity and war, and how men and women's everyday lives are affected by these global, colonial processes. Everyday contexts and lived experiences are therefore the sites where interlocked identities and oppressions manifest and where categories of inequality can be subverted. A truly decolonial intersectional approach is one that "examines how different methods or systems of oppression work together at different times and in various ways" (Gentry 2016, p. 27) through lived experiences but also as these relate to a global coloniality that is bigger than the gender dimension (Richter-Montpetit 2016, p. 108).

Consequently, intersectionality is a valuable research tool if used correctly – not as a multilayered compound understanding of cumulative oppressions and inequalities but as the exploration of open empirical categories that materialize from the dynamic interaction between the individual and institutional factors (Kantola and Nousiainen 2009). As Wibben (2016, p. 4) notes, although a more difficult approach than a singular one, it is more accurate with a greater potential for developing deeper insights. The emancipatory or decolonizing potential of intersectional peacebuilding research extends to the practices of resource allocation. For instance, in a Disarmament, Demobilization and Reintegration (DDR) resource allocation model, where the interdependent nature of gender and race is recognized, there is a better chance of resources reaching marginalized groups (such as "bushwives" who are not included in the narrow definition of a combatant instead of simply favoring male combatants).

Doing discourse and intersectional analysis together: a case study of sexual violence in the DRC

In this section I illustrate the decolonizing potential of the integrated use of critical-feminist discourse and intersectional analysis through the prism of the ground-breaking work of Maria Eriksson Baaz and Maria Stern (2013) on sexual violence committed by the Forces armées de la République démocratique du Congo (FARDC) in the DRC. In their book *Sexual Violence as a Weapon of War? Perceptions, Prescriptions, Problems in the Congo and Beyond*, the authors find that Congolese men rape due to a complex mix of cultural and political economy perceptions about masculinity, women as property and a sense of entitlement to sex as compensation for their loss of status as providers.

The authors made use of critical-feminist discourse analysis to challenge two interrelated narratives. The first is the liberal message of UNSCR 1820 (2008), which cast rape as a strategic and tactical weapon of war and recognizes rape and other forms of sexual violence as crimes against humanity. The second discourse of "Rape as a Logic of Barbaric African Wars" (Baaz and Stern 2016, p. 124) is founded on colonialist assumptions.

In the case of the former, Baaz and Stern highlight the way in which UNSCR 1820 became instrumental in establishing a rather cohesive and optimistic narrative on sexual violence during war, placed against the backdrop of the liberal peace. The discourse identified sexual violence during war as a clearly defined problem with a clear solution (Baaz and Stern 2013). Part of the success of this global policy discourse was because it did not look for causes in men's natural biology, but rather presented the problem as one of gendered militarization. Like in Rwanda, where genocide served as a deliberate strategy to target women in their role as child bearers to destroy the purity of the (enemy) ethnic group, sexual violence in the DRC came to be viewed as a tool of humiliation and intimidation of the enemy. In gendered terms, rape was framed as a tool to discipline women who did not adhere to the "essential" categories of manhood and womanhood, and in a way then becoming unworthy of protection (Baaz and Stern 2016, p. 121).

UNSCR formulated as part of the Women, Peace and Security architecture therefore become governance tools to curtail, among others, sexual violence. In this framing, the link with the liberal peace is quite strong, as holding offenders accountable implies that they (the rapists) are seen as rational, liberal subjects. Viewed from a postcolonial or decolonial perspective, the second discourse of "Rape as a Logic of Barbaric African Wars" is founded on principles of othering which depicts African and/or Congolese soldiers as deviants and barbarians (Baaz and Stern 2016, pp. 123–124). These radicalized stereotypes of African men leave very little room for more complex alternative explanations.

Baaz and Stern's fieldwork consisted of semi-structured interviews and participant observation (informal conversations). They conducted interviews with members of the FARDC, delving into their views about "good soldier" discourses and their understanding of gender roles in the military. Informal conversations with members of women's groups and testimonies at workshops also revealed women's own meaning-making regarding sexual violence and survival strategies (Baaz and Stern 2016, pp. 125–126). Using an intersectional lens that is cognizant of the historical and postcolonial context of the conflict and postwar situation in the DRC allowed the researchers to see men and women's war and peace experiences as heterogeneous. For instance, by not seeing soldiers as only soldiers, but also as human beings, fathers, husbands and breadwinners facilitated the uncovering of alternative narratives. Soldiers were relatively silent on issues of militarized masculinities and instead preferred to share their fears in relation to their dwindling agency as husbands and fathers (providers and protectors). Against this backdrop, rapes committed by soldiers were seen to be emasculating (Baaz and Stern 2016, p. 127). The informants provided little evidence that rape was ordered by commanders, dispelling the "myth" of strategic intent and also revealing the soldiers' suffering/victimhood in war as child soldiers as it intersected with their agency in violence.

In similar vein, making sense of women's diverse experiences was a complex and messy process. Global hype about war-related sexual violence has trapped the discourse in the language of women's victimhood. In this global climate of "spectacularizing" sexual violence, other issues such as violence perpetrated against family members and the mundane everyday insecurities of children's health and land politics are completely overlooked. Without an intersectional lens, the many "selves" of women as they find ways to survive will be missed. According to the Life & Peace Institute, the "sexualization" of rape has become big business. Baaz and Stern (2016, p. 128) discuss in this regard how the commercialization of rape manifests in the DRC. Women's organizations elevated the issue of sexual violence in order to access donor funding for rape survivor programs and women presented themselves as rape survivors in order to qualify for humanitarian aid (also see Jean-Bouchard 2013).

Baaz and Stern's field research revealed many ambivalences and contradictions, which led to a host of epistemological and ethico-political challenges. The disjuncture between the soldiers' and women's narratives and the researchers' preconceived notions of what they thought they knew before they entered the field not only led to a sense of surprise (Baaz and Stern 2016, p. 126), but also to shame about their complicity in the maintenance of dominant hegemonic discourses about sexual violence. But instead of seeing ambivalence and messiness as a hindrance, the researchers turned it into a productive "methodology of unease," described as a "journey [that] landed us in both familiar and strange places, and often the familiar became strange as we travelled down this road" (Baaz and Stern 2016, p. 117). There is therefore no recipe for trying to make sense of these often contradictory accounts. This case does show that the researcher's positionality can be productively employed as a resource rather than an obstacle.

What is made strange here are not only the discourses and the research methodologies, but also the very presumptions that underpin the research endeavor. Baaz and Stern carefully picked the stories (such as women's testimonies in policy reports) apart, focusing on how they were constructed and how gendered knowledge contributed to particular meanings associated with rape (Baaz and Stern 2016, p. 131). This enabled them to compare narratives on the ground with the implicit "takenforgrantedness" of global narratives, in a way also exposing the unspoken metaconsensus that exists among great powers about the liberal peace. Their unique approach to discourse analysis helped them to make familiar content strange – also in terms of gender. In this respect, they managed to draw attention to how different understandings of gender inform the construction of different discourses of gender and sexual violence locally and globally. So-called "normal" ethnographic research methods of semi-structured interviews and participant observation were made strange through their methodology of unease, where they employed constant reflexivity and critical engagement with complex identity categories.

Standard ethical considerations of doing no harm in the course of asking critical questions about the "rape industry" (Basini 2016, p. 163) were made strange through the use of an intersectional lens, allowing perspectives of women as more than just victims or passive recipients of donor money and of soldiers as human beings. The combination of traditional ethnographic research tools with poststructuralist and postcolonial feminist principles allowed for novel insights into the complex political economy landscape of resource and land conflict as well as governance issues that usually frame violence. Thus by "unfamiliarizing" both the content and methods of research, the foundation of the research enterprise, namely the authority and implicit power of the researcher as "knowing" and omnipotent, is uprooted.

Following from this, what is made "familiar" through this novel methodology of listening is the very fact that research subjects play an epistemic role in the research process. Their everyday insights about their own experiences of gender and sexual violence have a bearing on the personal and the particular, but also have the capacity to inform (and reframe) research questions and reorient the peacebuilding research agenda. Their bona fides therefore lie in their ability to break through the abstract, universalized nature of Western research practices. Baaz and Stern's approach has shown that by learning to deal with emotions and affective responses (one's visceral responses of fear, disgust, shame and ego) a researcher becomes better equipped to see the rapist or soldier in less one-dimensional terms, allowing the human and vulnerability to shine through. Such a methodology of subversion through making the familiar odd and taming the strange moreover underlines the uncomfortable fact (for Westerners) that rape is not always the biggest concern for women. Once we stop seeing just the familiar narrative, then it offers an opening for other issues and concerns to be considered as they intersect with matters of sexual violence.

What can we learn from reading the feminist scholarship, the methodological choices, the research strategies and the case study together, particularly for doing research on peacebuilding in African contexts? Baaz and Stern's research methodology (viewed through the lens of postcolonial and poststructuralist "strangemaking") offers methodological openings that are different from a bottom-up research methodology that privileges the "local." Instead their foregrounding of complex, intersected African identities and agencies at play in the research process makes for a truly inductive research process. The different ways of listening and feeling (methodology of unease) enabled them to hear both multiple and marginalized voices. Herein lies the roots of a truly decolonial methodology to African peacebuilding research, namely finding:

> ways in which African research can be better informed by local concerns and interpretations and at the same time, concurrently, for African experiences to be taken into account in general [global] theory-building, the structural racism of the global system not withstanding.
>
> (Oyewumi 2002, p. 1)

Conclusion

The focus of this chapter was on the intersection between feminist theory and feminist method and the consequences when they meet in the field of African peacebuilding. Oliver Richmond's concern that if the "theory" is difficult to define, it becomes difficult to "understand agency when we do not have the methodological tools to uncover it?" (Richmond 2011, p. 20) can thus be addressed by learning from feminist scholarship – by understanding first how the choice of feminist theory is connected to the strategy for its achievement; and second that all gender theory and methods are underpinned by an ethical framework of reflexivity. It is this "glue" that lays the foundation for a more coherent body of critical peacebuilding research. Failure to acknowledge this not only reinforces the fallacy that there is some kind of objective peace research standard "out there" but also legitimizes gender stereotypes as well as a masculine vantage point of hierarchy and detachment. Thus, the elements outlined in the chapter have the potential to uncover new dimensions of the subjecthood/agency of peacebuilders (both external and internal) as well as radically altering what we already know (making the strange familiar and the familiar strange). Baaz and Stern's methodology of unease is a case in point.

Poststructuralist and postcolonial feminist research on war and peace in the postcolony (with their integrated emphases on CDA and intersectionality) exposes the skewed assumptions about women and men's roles in violence and peace. These feminist theories, when placed in a particular conflict or post-conflict context, show us that gender is produced through fluid and multiple power relations and intersections with other identity markers. Identity categories are not innocent, neither is language – they are all imbued with bias. Therefore, if we want to develop more coherent critical peace theory, then the integrated nature of feminist theory and method (through discourse analysis and intersectionality) serves as a valuable starting point for decolonized peacebuilding research practices and theorizing. Feminist IR and security studies research alert us to the fact that fieldwork on peace and security in the postcolony in particular "is not a linear process of confirming hypotheses but rather a set of positionings through which researcher and researched negotiate with each other and in doing so establish their subjectivities" (Tami Jacoby in Basini 2016, p. 164). Herein lies the real transformative power of the feminist theory-research methodology nexus.

The implications of this approach for developing African-centered research methodologies – that are neither essentialist nor a clever copy-cat of Western experiments – are huge. Cultivating a sensitivity for theoretically informed methodologies holds value for research on Africa, where multiple conflicting discourses about Africa as the "dark continent" and "Africa Rising" cohabitate. The integrated ontological-epistemological lens offers the possibility of reconstructing African subjecthood in terms of complex mixes of subjectivity and victimhood – a heterogeneity that does not only see Africa as an interesting or deviant site where Western theory is applied, but also as the producer of situated forms of entangled knowledges.

Note

1 "Making the familiar strange and the strange familiar" is a phrase borrowed from Maria Eriksson Baaz and Maria Stern's work on sexual violence in the DRC (Baaz and Stern 2016, p. 134).

References

Acharya, Amitav and Barry Buzan (eds). (2010) *Non-Western International Relations Theory. Perspectives on and Beyond Asia.* London & New York: Routledge.

Autesserre, Séverine. (2014) *Peaceland. Conflict Resolution and the Everyday Politics of International Intervention.* Cambridge, MA: Cambridge University Press.

Barnett, M. (2006) "Building a Republican Peace," *International Security* 30 (4), pp. 87–112.

Basini, Helen. (2016) "'Doing No Harm'. Methodological and Ethical Challenges of Working with Women Associated with Fighting Forces/Ex-Combatants in Liberia," in Wibben, Annick T.R. (ed.). *Researching War. Feminist Methods, Ethics and Politics.* London & New York: Routledge, pp. 163–184.

Bercovitch, J. and R. Jackson. (2009) *Conflict Resolution in the Twenty-first Century. Principles, Methods, and Approaches.* Ann Arbor: The University of Michigan Press.

Bhabha, Homi K. (1994) *The Location of Culture.* London & New York: Routledge.

Boege, Volker, Brown, Anne, Clements, Kevin and Anna Nolan. (2008) *On Hybrid Political Orders and Emerging States: State Formation in the Context of 'Fragility'.* Berlin: Berghof Research Center for Constructive Conflict Management, Berghof Foundation.

Chandler, David. (2009) *Hollow Hegemony: Rethinking Global Politics, Power and Resistance.* New York: Pluto Press.

Chowdhry, Geeta and Sheila Nair (eds). (2004) *Power, Postcolonialism and International Relations: Reading Race, Gender and Class.* Oxford & New York: Routledge.

Crenshaw, Kimberle. (1991) "Mapping the Margins: Intersectionality, Identity Politics, and Violence against Women of Color," *Stanford Law Review* 43, pp. 1241–1299.

Duffield, Mark. (2001) *Global Governance and the New Wars: The Merging of Development and Security.* London: Zed Books.

Enloe, Cynthia. (1990) *Bananas, Beaches, and Bases: Making Feminist Sense of International Politics.* Berkeley: University of California Press.

Enloe, Cynthia. (2007) *Globalization & Militarism: Feminists Make the Link.* Lanham, MD: Rowman & Littlefield.

Eriksson Baaz, Maria and Maria Stern. (2013) *Sexual Violence as a Weapon of War? Perceptions, Prescriptions, Problems in the Congo and Beyond.* New York: Zed.

Eriksson Baaz, Maria and Maria Stern. (2016) "Researching Wartime Rape in the Democratic Republic of Congo. A Methodology of Unease," in Wibben, Annick T.R. (ed.), *Researching War. Feminist Methods, Ethics and Politics.* London & New York: Routledge, pp. 117–140.

Fairclough, Norman. (2012) "Critical Discourse Analysis," in Gee, James Paul and Michael Handford (eds), *The Routledge Handbook of Discourse Analysis.* London & New York: Routledge, pp. 9–20.

Foucault, Michel. (1972) *The Archaeology of Knowledge*. New York: Pantheon.
Gentry, Caron E. (2016) "Chechen Political Violence as Desperation. What Feminist Discourse Analysis Reveals," in Wibben, Annick T.R. (ed.), *Researching War. Feminist Methods, Ethics and Politics*. London & New York: Routledge, pp. 19–37.
Grosfoguel, Ramón. (2007) "The Epistemic Decolonial Turn. Beyond Political-Economy Paradigms," *Cultural Studies* 21 (2–3), pp. 211–223.
Haraway, Donna. (1988) "Situated Knowledges: The Science Question in Feminism and the Privilege of Partial Perspective," *Feminist Studies* 14 (3), pp. 575–599.
Harding, Sandra. (1986) *The Science Question in Feminism*. Ithaca, NY: Cornell University Press.
Heathershaw, J. (2008) "Unpacking the Liberal Peace: The Dividing and Merging of Peacebuilding Discourses," *Millennium – Journal of International Studies* 36 (3), pp. 597–621.
Hill Collins, Patricia. (1997) "Comment on Hekman's 'Truth and Method: Feminist Standpoint Theory Revisited': Where's the Power?" *Signs: Journal of Women in Culture and Society* 22 (2), pp. 375–381.
hooks, bell. (1984) *Feminist Theory: From Margin to Center*. Boston, MA: South End Press.
Hudson, Heidi. (2012) "A Double-Edged Sword of Peace? Reflections on the Tension between Representation and Protection in Gendering Liberal Peacebuilding," *International Peacekeeping* 19 (4), pp. 443–460.
Hudson, Heidi. (2016) "Decolonising Gender and Peacebuilding: Feminist Frontiers and Border Thinking in Africa," *Peacebuilding* 4 (2), pp. 194–209.
Hudson, Heidi. (2017) "The Power of Mixed Messages: Women, Peace, and Security Language in National Action Plans from Africa," *Africa Spectrum* 52 (3), pp. 3–29.
Jean-Bouchard, Évelyne. (2013) "Sexual Violence Issues in Eastern Congo: Normative Processes of Local and Global Co-constitutions." Presented at the International Studies Association Conference, San Francisco, CA.
Jennings, Kathleen. (2014) "Service, Sex, and Security: Gendered Peacekeeping Economies in Liberia and the Democratic Republic of the Congo," *Security Dialogue* 45 (4), pp. 313–330.
Kantola, Johanna and Kevät Nousiainen. (2009) "Institutionalizing Intersectionality in Europe," *International Feminist Journal of Politics* 11 (4), pp. 459–477.
Mac Ginty, Roger and A. Williams. (2009) *Conflict and Development*. London & New York: Routledge.
Mignolo, Walter D. (2011) *The Darker Side of Western Modernity: Decolonial Options*. Durham, NC & London: Duke University Press.
Milliken, Jennifer. (1999) "The Study of Discourse in International Relations: A Critique of Research and Methods," *European Journal of International Relations* 5 (2), pp. 225–254.
Mohanty, Chandra T. (1988), "Under Western Eyes: Feminist Scholarship and Colonial Discourses," *Feminist Review* 30, pp. 61–88.
Myrttinen, Henri, Jana Naujoks and Judy El-Bushra. (2014) *Rethinking Gender in Peacebuilding*. International Alert, http://www.internationalalert.org/ sites/default/files/Gender_RethinkingGenderPeacebuilding_EN_2014.pdf (accessed 13 October 2015).
Ndlovu-Gatsheni, Sabelo. (2014) "Global Coloniality and the Challenges of Creating African Futures," *Strategic Review for Southern Africa* 36 (2), pp. 187–188.

Oyewumi, Oyeronke. (2002) "Conceptualizing Gender: The Eurocentric Foundations of Feminist Concepts and the Challenge of African Epistemologies," *Jenda: A Journal of Culture and African Women Studies* 2 (1), pp. 1–9.
Paris, Roland. (2004) *At War's End. Building Peace after Civil Conflict*. Cambridge: Cambridge University Press.
Paris, Roland. (2010) "Saving Liberal Peacebuilding," *Review of International Studies* 36, pp. 337–365.
Paris, Roland and Timothy Sisk. (2009) "Introduction. Understanding the Contradictions of Postwar Statebuilding," in Paris, Roland and Timothy D. Sisk (eds), *The Dilemmas of Statebuilding. Confronting the Contradictions of Postwar Peace Operations*. London: Routledge, pp. 1–20.
Pratt, Nicola and Sophie Richter-Devroe. (2011) "Introduction: Critically Examining UNSCR 1325 on Women, Peace and Security," *International Feminist Journal of Politics* 13 (4), pp. 489–503.
Quijano, Aníbal. (2007) "Coloniality and Modernity," *Cultural Studies* 21 (2–3), pp. 168–178.
Reardon, Betty. (1993) *Women and Peace: Feminist Visions of Global Security*. Albany, NY: SUNY Press.
Richmond, Oliver P. (2009) "A Post-Liberal Peace: Eirinism and the Everyday," *Review of International Studies* 35 (3), pp. 557–580.
Richmond, Oliver P. (2010) "A Genealogy of Peace and Conflict Theory," in Richmond, Oliver P. (ed.), *Advances in Peacebuilding*. London: Palgrave Macmillan, pp. 14–38.
Richmond, Oliver P. (2011) *A Post-Liberal Peace*. London & New York: Routledge.
Richter-Montpetit, Melanie. (2016) "Militarized Masculinities, Women Torturers, and the Limits of Gender Analysis at Abu Ghraib," in Wibben, Annick T.R. (ed.), *Researching War. Feminist Methods, Ethics and Politics*. London & New York: Routledge, pp. 92–116.
Reynolds, Tracey. (2002) "Re-Thinking a Black Feminist Standpoint," *Ethnic and Racial Studies* 25 (4), pp. 591–606.
Roberts, D. (2008) "Post-Conflict Statebuilding and State Legitimacy: From Negative to Positive Peace?" *Development and Change* 39 (4), pp. 537–555.
Robinson, Fiona. (2011) *The Ethics of Care: A Feminist Approach to Human Security*. Philadelphia, PA: Temple University Press.
Sabaratnam, Meera. (2013) "Avatars of Eurocentrism in the Critique of the Liberal Peace," *Security Dialogue* 44 (3), pp. 259–278.
Sabaratnam, Meera. (2017) *Decolonising Intervention. International Statebuilding in Mozambique*. London: Rowman & Littlefield.
Said, Edward. (1978) *Orientalism*. New York: Pantheon Books.
Shepherd, Laura J. (2008) *Gender, Violence and Security: Discourse as Practice*. London & New York: Zed Books.
Shepherd, Laura J. (2011) "Sex, Security and Superhero(in)es: From 1325 to 1820 and Beyond," *International Feminist Journal of Politics* 13 (4), pp. 504–521.
Sylvester, Christine. (ed.). (2011) *Experiencing War*. London & New York: Routledge.
Sylvester, Christine. (2013a) "Experiencing the End and Afterlives of International Relations/Theory," *European Journal of International Relations* 19 (3), pp. 609–626.
Sylvester, Christine. (2013b) *War as Experience: Contributions from International Relations and Feminist Analysis*. London: Routledge.

The Santa Cruz Feminists of Color Collective. (2014). "Building on 'the Edge of Each Other's Battles': A Feminist of Color Multidimensional Lens," *Hypatia* 29 (1), pp. 23–40.

Tickner, J. Ann. (1992). *Gender in International Relations: Feminist Perspectives on Achieving Global Security.* New York: Columbia University Press.

Tickner, J. Ann. (1994) "Feminist Perspectives on Peace and World Security in the Post-Cold War Era," in Klare, M.T. (ed.), *Peace and World Studies: A Curriculum Guide.* Boulder, CO: Lynne Rienner, pp. 43–54.

Wibben, Annick T.R. (2016) "Introduction: Feminists Study War," in Wibben, Annick T.R. (ed.), *Researching War. Feminist Methods, Ethics and Politics.* London & New York: Routledge, pp. 1–16.

Wielenga, Cori. (2014) "'Lived' Identities in Rwanda: Beyond Ethnicity?" *Africa Insight* 44 (1), pp. 122–136.

3 A mission to civilize

The liberal idea of peacebuilding in Africa

Thomas Kwasi Tieku[1]

Introduction

Liberal peacebuilding, defined as international-led and engineered activities designed to create conditions to prevent resumption of violent conflict in a country, has been a buzzword in Africa, especially since the publication of the UN's Agenda for Peace in 1992 (UN 1992; Tieku 2012). The liberal approach to peacebuilding in Africa has enormous policy and normative implications but is rarely explored explicitly by scholars studying peacebuilding in Africa. In this chapter, I seek to show the policy and normative undercurrents of liberal peacebuilding in Africa, arguing that they have modernistic pretensions and are eerily close to the European civilizing mission in Africa during the 18th and 19th centuries. Put differently, liberal peacebuilding in contemporary Africa is a modernization or civilization project, but disguised as neutral technical exercise designed to create durable peace in fragile states and to prevent postwar countries from relapsing into violent conflict. While the "mission civilisatrice" 2.0 is built into all activities of peacebuilding including conflict prevention, management, resolution and postwar reconstruction, they are more pronounced in the postwar reconstruction activities. Based on these observations, I call for a research agenda in the field of African peacebuilding that do not only seek to uncover and understand the civilizing undercurrents of their areas of research but in Kwame Nkrumah's (1976, p. 18) words "actively combat neo-colonialism in whatever guise it may appear."

I am urging researchers in the field of African peacebuilding to take a normative approach to studying their subject primarily because researchers are trained to be technicians and to remain silent on the normative motivations behind our line of inquiry. It will, however, be a big mistake for African researchers of peacebuilding to put their normative cards in the closet. Not only will it be helpful for African peacebuilding researchers to put their normative cards on the table, it is the sincerest way to do research. As the Canadian political scientist Robert Cox famously put it, "every academic enterprise is always for someone and has some purpose" (Cox 1986). Cox entreated social science scholars to always ask two critical questions, namely who is our research for and what purpose does it serve? If the Coxian approach to social sciences scholarship is applied to African peacebuilding, it will become apparently clear that peacebuilding in Africa has generally been a political engineering endeavor. As such, any researcher who approaches peacebuilding in Africa as a technical activity will miss key elements of the peacebuilding enterprise. The researcher will likely ignore the political intentions of African peacebuilding. Moreover, it is essential for researchers to feel free to discuss their normative goals for exploring specific topics, puzzles and questions in the field of African peacebuilding. This is important given that genuine neutrality is rarely displayed in peacebuilding spaces.

In fact, the history of peacebuilding on the African continent in the last half century is a history of social construction of postwar societies to reflect the image of external powers. Liberal peacebuilding which is currently the most dominant model of peacebuilding in Africa, carries many residues from the civilizing mission of the 18th and 19th centuries (Zanotti 2011; Richmond and Mac Ginty 2015). The civilizing logic behind liberal peacebuilding practices, as shown in the sections below, is reflected in peacebuilders' sustained efforts to replicate the Weberian and Westphalian state architecture in Africa (state building in peacebuilding language); spread values and beliefs held by external interveners and their donors (good governance in peacebuilding parlance); reconstruct indigenous practices, identities and rules to reflect behavioral traits similar to those in donor states (post-conflict reconstruction in peacebuilding lexicon); and stimulate local people's taste for merchandized goods produced in donor countries (economic reforms in peacebuilding terms).

The civilizing intentions of liberal peacebuilding activities in Africa have been acknowledged by African policymakers. This is reflected in African Union's (AU's) main policy documents, declarations, resolutions, decisions and pronouncements on peacebuilding(African Union 2001, 2006; Akonor 2010). While the view that peacebuilding is a political exercise rather than a technical exercise is widely shared among the African political class, it is less acknowledged explicitly in the work of African scholars studying peacebuilding. Failure by African intellectuals in the field of peacebuilding to probe deeply into the normative assumptions and policy implications of their research put them in a position where they may unwittingly be enhancing the civilizing intentions of the liberal peacebuilding project.

The rest of the chapter is organized into four sections. The section following this introduction provides a review of the African peacebuilding literature with a view to showing its normative underpinnings and weaknesses. The next section shows the civilizing mission of the dominant peacebuilding practices in Africa. The penultimate section demonstrates that the civilizing mission of current peacebuilding activities is widely acknowledged by the African leadership class and policymakers. The conclusion returns to the central argument of the chapter and its implications for scholarship.

The African peacebuilding landscape

The techniques used by the international community, especially the UN and Western powers, to create conditions they consider necessary for peace to emerge and endure in Africa range from the continuum of conflict prevention to postwar construction activities (Jen 2012; Kreps 2007; Manning 2003). The full spectrum of peacebuilding activities, including conflict prevention, management, resolution and postwar construction, are designed to transform political, economic and social institutions in war-ravaged states into representative-democratic states, market-oriented economies and rule-based and socially liberal countries (Curtis and Dzinesa 2012; Mac Ginty 2013; Richmond and Mac Ginty 2015). Inspired by the modernization project, liberal peacebuilders see and treat democratization, market liberalization and liberal social values such as civil and political rights as the most effective and 'standard interventions' for creating durable peace and development in war-prone states (Paris 2002, 2010; Richmond 2009). One of the preferred techniques employed by liberal peacebuilders in war-torn countries in Africa is the deployment of peacekeeping troops to maintain externally engineered and brokered peace agreements signed by national political elites of war-affected African states (Williams and Mengistu 2015). The deployment of troops usually occurs in one of four ways: first, after internationally engineered negotiation of a ceasefire agreement as it was the case in the Democratic Republic of Congo; second, after African-led and internationally backed mediation of a conflict as happened in South Sudan; third, after a formal colonial-master has created a semblance of peace as it was the case in Mali; and finally, else a combination of any of the three options. Whether codenamed peacekeeping operations or peace support operations or peace enforcement, these liberal peace missions are usually authorized by the United Nations Security Council under chapter VI or VII before deployment as occurred in Cote d' Ivoire or after intervention by African actors as was the case in Sierra Leone.

In most cases, liberal peace missions are undertaken solely by the UN with the political and financial support of major global powers and international financial institutions, or jointly embarked upon by the UN and African regional organizations. The UN acted as a main driver of liberal peace in places such as Sierra Leone, Liberia, Mozambique and the Democratic Republic of Congo. The UN and the AU jointly carried the liberal peace banner in the Darfur region of Sudan. In some cases, the UN took over or absorbed an African-led mission, in part to deepen and sharpen the liberal elements of the peace as seen in Burundi after the Arusha peace accord. Besides the UN, African regional organizations have been used indirectly to promote and enforce liberal peace (Murithi 2009). The indirect form of liberal peacebuilding often occurs when non-African actors use control over resources such as payment of allowances and provision of logistics to African-deployed troops to promote democratization, market liberalization and liberal social values in war-torn African states. The African Union Mission in Somalia (AMISOM) is a good illustration of peacekeeping that has become an instrument of indirect form of liberal peacebuilding.

Peacekeeping as a tool for creating durable liberal peace has come under a barrage of criticisms. These criticisms can be categorized into four groups: group one criticisms focus on challenges inherent in peacekeeping rules of engagement, group two criticisms are those identified during the practice of traditional peacekeeping, group three critiques touch on mismatch between peacekeeping objectives and resources deployed and the last batch of limitations revolve around the unintended insecurity that the deployment of troops create for so-called fragile societies.[2] As a result of these limitations associated with classic peacekeeping, the new buzz-phrase for liberal peacebuilders is peace support operations which involve active conflict and the use of armed peacekeepers to protect civilians. But as the experience of regional organizations such as the AU's intervention in Somalia showed, the casualty rate is too high for the risk-averse liberal internationalists who are the main backers of liberal peacebuilding. For instance, it is estimated that between 2001 and 2017, "the number of casualties among African troops in (AU) peace support operations exceeded the combined casualties in United Nations peacekeeping missions in the last 70 years" (African Union 2017, p. 2, United Nations 2017). The high casualties involved in peace support operations make it hard for liberal peacebuilders to embrace the idea wholeheartedly. In addition, research findings such as those documented by Michael Kobi, David Kellen and Eyal Ben-Ari (2009) showed that peace support operations have their own weaknesses including limitations in the areas of strategic planning, operations and tactical deployment of troops.

The unintended consequences of the presence of large contingent of international actors in conflict-affected countries have made both peacekeeping and peace support operations ineffective liberal peacebuilding tool. Research demonstrated that the mere presence of the UN and peacekeepers generate its own political economy that feeds existing violence and/or creates new war (Ahmad 2012; Jan 2001). Jan's (2001, p. 72) research showed that the unintended consequences of actions taken by UNOSOM II and the mere presence of UNOSOM II fueled the post-Barre civil wars in Somalia. Some of the activities led to stagnation of the Somali economy which undermined peace efforts, while others created an underground economy that fed warlords making it difficult for the warring parties to lay down their arms. Other UNOSOM II activities created new actors who were more interested in the continuation of violence because they build their careers and businesses around the continuation of the war. Similar observations were made by Ahmad in 2012. In Ahmad's view the UN peacekeeping budget, which is currently more than $6 billion annually, has become a budget for war rather than a budget for peace.

Finally, liberal peacebuilders often find it difficult to adopt peace operations to local needs and concerns (Cali and Cook 2003; Donais 2009). Many managers of liberal peacekeepers approach their work from an elitist perspective. It is therefore hard for them to spot local dangers and address challenges that face ordinary people. The elitist approach might in part explain why the presence of international peacekeepers increases the possibility of structural violence including gender-based violence and sexual exploitation and abuse of vulnerable groups in society. It can also reinforce patriarchy in many warzones. The negative unintended consequences peace operations have been used by governments in Burundi and DRC to push for an end to UN peacekeeping operation in their countries.

In recent years, liberal peacebuilders led by the UN and Western powers have turned their attention to conflict prevention as a tool to create conditions for durable peace in the so-called fragile African states. The conflict prevention turn, which is more apparent in the rhetoric of liberal peacebuilders than in actual practice, led to the creation of UN Mediation Support Unit in 2006, the relaunch in 2012 of the UN Peacemaker and the establishment of a High-Level Advisory Board on Mediation to advice the UN Secretary-General on mediation initiatives and to provide political support to specific UN-supported mediation activities around the world.

Despite the popularity of mediation as a conflict prevention tool, it can at best prevent the outbreak of war or contribute to the management of violent conflict in the short term. Successful mediation activities on the African continent have often resulted in political patchwork and elite political accommodation for a brief period. Classic cases include the creation of transitional governments in Burundi from 2001 to 2005 and the establishment of coalition government in Kenya from 2008 to 2013 and in Zimbabwe from 2009 to 2013. As subsequent events in Burundi, Kenya and Zimbabwe have shown, the mediated approach to peacebuilding has the potential to create necessary conditions to prevent the outbreak and spread of armed conflict, albeit for brief periods, but cannot create an optimal environment for durable peace to emerge.

Moreover, mediation as a tool for building liberal peace has become popular in recent years, yet it has a number of challenges and limitations. In the interest of brevity, only two of these limitations will be highlighted. Mediation has been criticized from a gendered lens. Mediation processes tend to be gendered in the sense that women are usually given specific roles; their inclusion tends to be circumscribed and ghettoized. For the most part, women are given pink jobs and excluded from high powered political negotiations. Puechguirbal's (2012) work showed that women are defined primarily as victims in peace processes yet excluded from participating in official peace negotiations. Even elite women are often excluded from the actual peace negotiations and the drafting of peace agreements where resources are allocated. Pankhurst's (2010) research uncovered that it is not only mediation roles that are clearly aligned along gender lines, but so are needs.

It is perhaps unsurprising that political arrangements that come out of mediated settlements tend to favor men. This was the case in the transitional government of Burundi that the Arusha peace process established. It was also evident in the political arrangement that came out of the Darfur negotiation and the comprehensive peace agreement South Sudan and Sudan signed. Key positions including the president, vice-president and key ministerial portfolios were all given to men. Indeed, those who end up holding major political positions in most mediated conflict on the African continent are men. Women who were included in these governments that emerged after peace talk were confined to the role of managing social issues or low political issues while their male counterparts occupied the primetime political positions.

The limitation of peacekeeping and conflict meditation as tools to create durable peace is perhaps why the liberal peacebuilding agenda has often been advanced through the post-conflict reconstruction activities. The UN and other liberal peacebuilders have packaged activities such as state building, Demilitarization, Demobilization and Reintegration (DDR), security sector reforms, democratization, rule of law, liberal social values and economic liberalization as the key to preventing war-ravaged societies from relapsing into violent conflict. Manning (2003) showed that the biggest challenge facing DDR processes in postwar societies is the reintegration of demobilized soldiers and displaced persons into local communities together with the reintegration of opposition-controlled territory and assets into a unified state administration.

These measures are considered by liberal peacebuilders as offering the sufficient condition for peace to emerge but as will be discussed in the next section, these post-conflict reconstruction activities are not just meant to create peace, they are designed to reconstruct conflict-affected societies to reflect the interests, preference and desires of external powers (Zanotti 2011). In other words, the liberal peacebuilding process as reflected in the post-conflict reconstruction activities is a modernization project. Post-conflict reconstruction activities are similar to the 19th-century civilizing mission but disguised as a neutral technical project designed to prevent fragile states and postwar states from relapsing into violent conflict. This is certainly the case in Africa where most postwar societies have been made to implement peacebuilding activities reminiscent of the modernization project and eerily close to the civilizing mission. The next section explores this issue further.

The civilizing undercurrents of liberal peacebuilding in Africa

The modernization and civilization intentions are apparent in liberal peacebuilders' insistence and push to turn every postwar African society into a Westphalian state, representative democracy, market-led economy and model Western-style state. The obsession by liberal peacebuilders with the Westphalia state model is very much evident in the amount of resources and energy expended to keep failing states such as Somalia intact and the zero tolerance for irredentism. As Grovogui (2002) pointed out, international support for Africa including peacebuilding across the African continent in the last 50 years has been premised on the creation and/or strengthening of the "Weberian and Westphalian commonsense." The Weberian and Westphalian commonsense has been promoted by international organizations such as the UN, World Bank and Africa's own regional and continental organizations. As Richmond (2003) observed, the state is the sole unit given primary consideration in liberal peacebuilding activities. The overemphasis on the state at the expense of local levels of governance made Richmond advocate for a hybridized model of peacebuilding where liberal notions of representation will be combined with culturally appropriate mechanisms to generate what he termed as "liberal-local hybrid."

Peacebuilding as a state building exercise has focused primarily on legitimizing colonial boundaries, institutionalizing central government's control over these territories and developing norms to protect the territorial integrity of colonially demarcated areas. Thus, as even liberal internationalists such as Paris (2002) agreed, liberal peacebuilders are interested in creating a particular model of state organization based on liberal democracy and market-oriented economics. Paris admitted that the application of liberal peacebuilding resembles civilizing missions and exploitation during the colonial period, but rejected the idea that liberal peacebuilders are interested in or consciously try to exploit and/or civilize postwar societies. In his view, liberal peacebuilding is driven by ideology rather than the logic of exploitation and civilization.

For some scholars, it is the Agenda for Peace that turned peacebuilding into a state building enterprise. As Cali and Cook (2003, p. 235) noted, most scholars and practitioners "rely on the currency given the term by Boutros Ghali's 1992 *Agenda for Peace*, which viewed peacebuilding as post conflict social and political reconstruction activities [...or] statebuilding in societies emerging from armed conflict." Attempts to replicate the Western state system has been approached in different ways, including military intervention to prevent the collapse of central government as in the case of Somalia, mediation aimed at creating space for elite accommodation as in the examples of Kenya and Zimbabwe in 2008, the establishment of a new political path as in the instances of Liberia and Sierra Leone and institutional design to ensure that different ethnic groups coexist in a nonviolent way as in the situation of Burundi in the early 2000s. Little opportunity, Clapham Christopher (2012) showed, is given to those who want to imagine alternative political systems which may perhaps be more suitable to the societies where peace is being built. Through liberal peacebuilding exercises and other instruments, African elites have been socialized into believing that the Westphalia model is the only game in town. Many of them seem to have no mental space to consider alternative governance model.

The language of good governance has been used by liberal peacebuilders, especially in the last 20 years, to promote Western representative style of democracy (Richmond and Mac Ginty 2015; Zanotti 2011). Good governance in the African peacebuilding context focused on ensuring neoliberal efficiency in public services, the rule of law in contract management, effective adjudication of disputes usually between elites, respect for political and civil rights, pluralistic and representational form of government and participation by civil society groups and NGOs in political governance (World Bank 2002). These measures are designed to prevent the outbreak of violence (negative peace), enhance the capacity of holders of state authority to exercise effective coercive control over critical mass of people and creating the space for elites to share power and state resources. These elements are threads that hold together the Westphalia model or Western style of government. The goal of liberal peacebuilders is to plant these liberal ideas in postwar societies. The language of good governance provides a good cover for liberal peacebuilders to institutionalize Western imaginary ideas of good political arrangements in postwar states without any major protestation. As Zanotti informs us "good governance is a capacious concept that creates consensus around matters that would otherwise remain politically controversial" (Zanotti 2011, p. 56).

In the last ten years, the idea of peacebuilding has been associated closely with the new fad in town, post-conflict reconstruction. Most donor states have used the language of post-conflict reconstruction to support peacebuilding activities that they have been supporting since the 1960s. These activities have included conflict resolution, DDR, security sector reforms and the new religion called democracy. These reconstruction activities are designed to replace traditional and customary practices in societies emerging from conflict. In other words, postwar reconstruction programs are not fundamentally different from modernization projects. Indeed, countries that have emerged from war have almost identical national institutions. The similarities of the political institutions in Liberia, Sierra Leone, Mozambique and Burundi, just to mention a few, are not just sheer coincidence. In many instances these political institutions are more in tune with societies in donor states than those in postwar countries. As Cali and Cook aptly pointed out,

> [t]he concept of peacebuilding now means not only keeping former enemies from going back to war... the international community is more bound and determined to create democratic political regimes in postwar settings, recreating the core institutions of state and society largely along Western lines of thought.
>
> (Cali and Cook 2002, p. 233)

Liberal peacebuilding in the last 30 years has focused on restructuring the economies of countries emerging from war (Curtis and Dzinesa 2012; Mac Ginty 2013; Richmond and Mac Ginty 2015; Zanotti 2011). Liberal peacebuilders have created room for the adoption of neoliberal economic policies such as the removal of social safety nets, promotion of imports from donor countries and the creation of a viable private sector-led economic development. In some cases, liberal peacebuilders focus on discouraging governments in postwar states from pursuing policies that will enhance local production and in other cases they refrained from providing crucial material support for lifestyles and practices that international interveners consider anti-modern or traditional. As Sabaratnam's (2017) research in Mozambique showed, liberal peacebuilders not only refrained from supporting policies that would provide crucial material support for peasant and subsistence farming lifestyles, international interveners tried to make postwar states dependent on goods and services produced in donor countries. The result is that most of the African countries that have emerged from wars have tended to be outward looking and have exposed themselves to global economic competition. Most of these countries have removed a number of measures such as social safety net, agricultural support and state-funded education that protected vulnerable groups. Almost all of them have ended up changing the taste of a significant number of their populations from consuming locally produced goods in favor of consumption of foreign imported goods and services. The consumption pattern of people in postwar societies such as Burundi, Sierra Leone, Mozambique and Angola now mirrors that of the citizens of donor states. In some instances, the new consumption pattern reproduces conflict as Aisha Ahmad's (2012) research uncovered in Somalia. Thus, the liberal peacebuilding experiment in postwar African countries has gone beyond what has been possible with structural adjustment policies in countries that have not experienced large scale war. The view that liberal peacebuilders are reconstructing societies in postwar African states is widely shared among the African political class, as the next section shows.

The civilizing mission of liberal peacebuilding and the African political class

The AU, the main representative body of the African political class, has recognized how the current liberal peacebuilding model makes it easy for external actors, especially donors, to reconstruct the identities of people in societies emerging from war. It developed the Post-Conflict Reconstruction and Development policy specifically to discourage liberal peacebuilders from turning war-torn African societies into photocopies of outside cultures (African Union 2006). The document sought to provide benchmarks designed to help AU institutions, African governments and local government officials to determine the content of peacebuilding projects and to assess, implement, monitor and evaluate them in such a way as to reinforce core local identities and values. Attached to the postwar reconstruction policy is a "flexible template" that its framers think—perhaps naively—will encourage peacebuilders to move beyond civilizing goals to "pave the way for growth and regeneration in countries and regions emerging from conflict" (African Union 2006).

The postwar reconstruction template is also designed to dissuade peacebuilders from prioritizing some aspects of the reconstruction processes at the expense of others. A major complaint from civil society groups during AU–civil society consultations on postwar reconstruction is that many donors and practitioners tend to favor aspects of reconstruction that will generate publicity and media attention to the exclusion of other important, but less visible, areas. Often the ones that are neglected, such as psychological healing, retraining of former combatants and creation of sustainable jobs, are precisely those most needed by the people recovering from war. To address this, the postwar reconstruction policy provided benchmarks for measuring successes and failures of peacebuilding. Accordingly, every postwar peacebuilding process in Africa must strive to free the so-called ordinary Africans from "want and fear." Although the African Peace and Security Architecture's (APSA's) postwar reconstruction policy is general and imprecise, and at times contradictory and incoherent, its drafters made an excellent move by establishing linkages between the ideas in it and extant AU policies. The security elements were thus embedded in human security ideas outlined in the African Non-Aggression and Common Defence and Security Pact (African Union 2005b).

The political governance element drew extensively from the AU Declaration on Political, Economic and Corporate Governance of 2002, which contains the textbook conception of political governance (African Union 2002b). The AU has observed the move to make postwar African states mirror images of the Westphalian and the Weberian model. The model, promoted through institutions of the UN focused primarily on legitimizing colonial boundaries, institutionalizing the colonial regimes' control over these territories and developing norms to protect the territorial integrity of colonially demarcated areas. The AU developed the APSA in 2002 in part to take control and to shape the nature of the state building process (African Union 2002a). As a result, one of the things that the APSA tried to do is to change the traditional notion of state sovereignty by seeking to protect African states from military intervention by non-African states while leaving room for the AU to intervene in the internal affairs of African states with or without the consent of the target country.

The intervention can take different forms, including mediation, as in the case of Kenya in 2008, and suspension from participation in the activities of African international organizations, as in the case of Mauritania in 2009. In some cases, intervention can take the form of rebuke and suspension of AU membership, as in the case of Cote d'Ivoire in 2011; and as a last resort, military intervention, as in the case of Comoros Island in 2008. Conditions for military intervention are detailed in Article 4(h) of the Constitutive Act of the AU. The threshold conditions established by the APSA, as Maxi Schoeman pointed out, "go 'beyond' the provision made for intervention in the internal affairs of a country in the UN Charter" (Schoeman 2002). Thomas G. Weiss (2004) also argued that the threshold conditions are lower than those outlined in legal codes of other region-wide institutions. The conditional sovereignty introduced into Africa's international subsystem by the APSA is predicated on *ubuntu's* (i.e. collectivist) view of persons and communities.

The APSA sought to change the use of military intervention by external actors to further the modernization project by redefining humanitarian intervention in Responsibility to Protect (R2P) terms. Various APSA documents, in particular the *Ezulwini* Consensus, reiterated support for the three pillars of R2P: the responsibility of states to protect their citizens; the responsibility of the international community to help states protect their citizens; and the responsibility of the international community to protect citizens of states that are unable or unwilling to protect their citizens (African Union 2005a).

The APSA created the space for African political class to take ownership for intervention and state building by introducing three principles that seek to undercut the influence of liberal peacebuilders. First, the APSA shifted to regional organizations the power to decide when, where, and how to intervene, contrary to the argument put forth by liberal internationalists who see the UN as the best body to make those decisions (Weiss 2004; Welsh 2013). As noted in the *Ezulwini* document, "the General Assembly and the Security Council are often far from the scenes of conflicts and may not be in a position to undertake effectively a proper appreciation of the nature and development of conflict situations" (African Union 2005a). The collective wisdom of members of the AU is that regional organizations are the institutions best placed to make the appropriate assessment and should be "empowered to take actions in this regard."

Second, the APSA de-linked R2P from regime change. According to the *Ezulwini* document, even though "it is important to reiterate the obligation of states to protect their citizens, this should not be used as a pretext to undermine the sovereignty, independence and territorial integrity of states." Third, while the APSA affirmed the UN Charter's insistence that intervention by regional organizations "should be with the approval of the Security Council," an interesting and subtle qualification was inserted into the *Ezulwini* Consensus that effectively makes it possible for regional organizations to seek the Security Council's approval after interventions. The *Ezulwini* Consensus also indicated that the UN should "assume responsibility for financing such operations." The language used here in the document is meant to encourage the UN to take fund but not to dictate the terms of peace imposed by regional organizations. This regional-led approach to peace and security was introduced into the African international order when the Economic Community of West African States (ECOWAS) Ceasefire Monitoring Group (ECOMOG) intervened in Liberia and Sierra Leone in the 1990s. The AU copied it when it intervened in Burundi in 2003. It has since become the standard approach to intervention by the AU. The AU has socialized the African political class and even the UN leadership to accept it as the new way of international intervention.

The African political class recognized that the traditional military approach to African security is part of the problem, and therefore sought to shift security discourses and practices in Africa away from traditional state military concerns to quotidian concerns of individual Africans. The APSA therefore redefined security as a multidimensional phenomenon, going beyond military considerations to include economic, political and social aspects of the individual, the family and the society. In the words of a key APSA document called the Conference on Security, Stability, Development, and Co-operation in Africa (the CSSDCA) "The concept of security must embrace all aspects of society... [and the] security of a nation must be based on the security of the life of the individual citizens to live in peace and to satisfy basic needs" (African Union 2002b).

To ensure that peacebuilders in Africa are held accountable, the APSA sought to introduce a third-party oversight into peacebuilding processes to manage the current system, which has no institutional mechanisms for liberal peacebuilders to render accounts of their performance. The APSA is trying to position the AU Commission as the third-party oversight institutions of peacebuilding in postwar African societies. If successful, it will serve as a major correction to the current system, which relies primarily on the court of public opinion. Without a third-party obligation and oversight, it has become too easy for peacebuilders to use financial muscle, ideational power and institutional reforms to reconstruct postwar states in their image. Finally, the APSA offered institutional mechanisms for parties in postwar societies to resolve disputes with peacebuilders. Disagreements are endemic in the peacebuilding processes and in countries emerging from war; the current arrangement does not have enough institutionalized mechanisms for parties to resolve their differences. The APSA tried to link reconstruction to other rules, regulations, norms and mechanisms the AU has developed to manage such disputes. These include requiring member states to inform the AU Commission of key postwar activities such as the holding of elections and to invite the AU Electoral Assistant Unit to observe elections.

Conclusion

Postwar African states have been subjected to liberal peacebuilding. In a Gramscian sense (See Gramsci 1971), liberal peacebuilding has acquired hegemonic status in postwar societies in Africa. Yet, liberal peacebuilding is neither a value neutral exercise nor a purely technical project. Rather, it is a political enterprise designed to change postwar African societies to reflect the image and interests of peacebuilders. The civilizing intentions behind liberal peacebuilding are clearly reflected in the modernization activities pursued by liberal peacebuilders, the unflinching quest to replace indigenous governance systems in Africa with the Western style of representative government in the form of presidential or parliamentary governments, the incessant desire to replace traditional and customary rules in postwar states with statutes, case laws and adversarial judicial processes similar to those existing in donor states, the proselytizing approach to spreading values and ideas held by external interveners and their donors in postwar African states and the ceaseless push to stimulate the taste of citizens of postwar states for merchandized goods produced in donor countries. Put together, there is little doubt that liberal peacebuilding is similar to the 18th- and 19th-century civilizing missions in Africa. Both set out to reconstruct Africans in the image of external interveners.

Curiously, there have not been enough attempts to push back or fight this subtle and in some cases overt ideological and cultural colonization through liberal peacebuilding. Like the 19th century when some Africans became transmission belts for colonization, some policymakers in Africa are promoting the liberal peacebuilding agenda. Some of the policies and programs implemented by postwar African governments either independently or in collaboration with non-African actors are disconcertingly close to liberal peacebuilding activities. The liberalization of postwar economies and politics all carries significant elements of liberal peacebuilding. Other Africans are using the vehicle of NGOs to advance the liberal peacebuilding agenda.

However, some policymakers have noted the damaging impacts of liberal peacebuilding civilizing work and have started to fight back. The AU recognized the political nature of liberal peacebuilding in the early 2000s and developed the Post-Conflict Reconstruction and Development policy in 2006 to lead the effort to delink liberal peacebuilding from the civilizing mission. The postwar reconstruction document came with a template designed to deter liberal peacebuilders from engaging in civilizing activities and to discourage funders of peacebuilding from prioritizing aspects of reconstruction processes that serve interests of donors at the expense of those that will benefit postwar societies the most.

It is however unclear if the AU Commission, which is the implementing agency of the AU postwar reconstruction ideas, will be able to raise the necessary resources and develop the required capacity to translate the ambitious postwar strategy into reality, given that the AU is financially dependent on the generosity of major supporters of liberal peacebuilding. The postwar plan came without any additional institutional support for the overburdened AU Commission. The AU Department of Political Affairs and the Department of Peace and Security does not seem well equipped to carry out their original mandate together with the implementation of the postwar projects. Neither does the capacity to implement the policy exist in many African states nor in most African civil societies. The few African states that have expertise and resources seem reluctant to take on the leadership role in the postwar reconstruction process. It seems unavoidable that resources and capacity challenges will oblige the AU Commission to neglect some aspects of the reconstitution processes. This notwithstanding, the AU Commission may be unwilling to use its institutional and financial challenges as an excuse to do nothing. The Commission will need the support of well-meaning scholars in the field of peacebuilding if the Pan-African body were to be successful in stripping off the colonial and civilizing imprints on liberal peacebuilding.

Notes

1 I am grateful to Maame Adwoa Nyame Sam of University of Ghana and Benjamin Douglas Drummond of King's University College at Western University for research assistance. The views expressed here and any error in the article are solely mine.

2 These limitations are discussed in detail in academic works which are widely available and there is no need to rehash all of them. For example, see Berdal (1996); Kobi and Kellen (2009); Yamashita (2012); Boutellis and Williams (2013).

References

African Union. (2001) *The Constitutive Act*. Addis Ababa.

———. (2002a) *The Protocol Relating to the Establishment of the Peace and Security Council of the African Union*, adopted by the 1st Ordinary Session of the Assembly of the African Union. African Union: Durban: July 9. http://www.peaceau.org/uploads/psc-protocol-en.pdf.

———. (2002b) The Conference on *Security, Stability, Development, and Co-operation in Africa* (CSSDCA). African Union: Addis Ababa.

———. (2005a) *The Ezuline Consensus: The Common African Position on the Proposed Reform of the United Nations*. New York: United Nations. http://www.africa-union.org/News_Events/Calendar_of_%20Events/7th%20extra%20ordinary%20session%20ECL/Ext%20EXCL2%20VII%20Report.doc.

———. (2005b) *The African Union Non-Aggression and Common Defence Pact*, adopted by the Fourth Ordinary Session of the Assembly. Abuja, January 31, 2005. https://au.int/sites/default/files/treaties/37292-treaty-0031_-_african_union_non-aggression_and_common_defence_pact_e.pdf. Accessed July 2, 2015.

———. (2006) *Report on the Elaboration of a Framework Document on Post Conflict Reconstruction and Development (PCRD)*. Banjul: The Gambia.

———. (2017) Statement of Ambassador Smail Chergui Commissioner for Peace and Security to the United Nations Security Council. African Union: Addis Ababa.

Ahmad, Aisha. (2012) "Agenda for Peace or Budget for War? Evaluating the Economic Impact of International Intervention in Somalia," *International Journal* 67(2), pp. 313–331.

Akonor, Kwame. (2010) "Assessing the African Union's Right of Humanitarian Intervention," *Criminal Justice Ethics* 29(2), pp. 157–173.

Berdal, Mats R. (1996) "The United Nations System of Command and Control of Peacekeeping Operations," *The International Spectator* 31(1), pp. 13–24. January–March. Published online April 2008.

Boutellis, Arthur, and Williams, Paul D. (2013) *Peace Operations, the African Union, and the United Nations: Toward More Effective Partnerships*. New York: International Peace Institute.

Cali, Charles T., and Cook, Susan E. (2003) "On Democratization and Peacebuilding," *Global Governance* 9(2), pp. 233–246.

Christopher, Clapham. (2012) "Peacebuilding Without a State: The Somali Experience," in Devon Curtis and Gwinyayi A. Dzinesa (Eds), *Peacebuilding, Power, and Politics in Africa*, Athens: Ohio University Press, pp. 295–309.

Cox, Robert. (1986) 'Social Forces, States and World Orders: Beyond International Relations Theory', in Robert Keohane (Ed), *Neorealism and Its Critics*. New York: Columbia University Press 1986, pp. 207–208.

Curtis, Devon, and Dzinesa, Gwinyayi A. (Eds). (2012) *Peacebuilding, Power, and Politics in Africa*. Athens: Ohio University Press.

Donais, Timothy. (2009) "Empowerment of Imposition? Dilemmas of Local Ownership in Post-Conflict Peacebuilding Processes." *Peace & Change* 34(1), pp. 3–26.

Gramsci, Antonio. (1971). *Selections from the Prison Notebooks of Antonio Gramsci*. New York: International Publishers.

Grovogui, Siba. (2002) "Regimes of Sovereignty: International Morality and the African Condition," *European Journal of International Relations* 8(3), pp. 315–338.

Jan, Ameen. (2001) "Somalia: Building Sovereignty or Restoring Peace?," in Elizabeth M. Cousens and Chetan Kumar (Eds), *Peacebuilding as Politics: Cultivating Peace in Fragile Societies*, Boulder, CO: Lynne Rienner, pp. 68–69.

Jen, Abou. (2012) *Peacebuilding in the African Union: Law, Philosophy and Practice*. Cambridge: Cambridge University Press.

Kobi, Michael, and Kellen, David. (2009) "Cultural Intelligence for Peace Support Operations in the New Era of Warfare," in Kobi Michael, David Kellen, and Eyal Ben-Ari (Eds), *The Transformation of the World of War and Peace Support Operations*. Westport, CT: Praeger, pp. 157–172.

Kobi, Michael, Kellen, David, and Ben-Ari, Eyal (Eds). (2009) *The Transformation of the World of War and Peace Support Operations*. Westport, CT: Praeger.

Kreps, Sarah E. (2007) "The United Nations-African Union Mission in Darfur: Implications and Prospects for Success," *African Security Review* 16(4), pp. 66–79.

Mac Ginty, Roger (Ed.). (2013) *Routledge Handbook of Peacebuilding*. London and New York: Routledge.

Manning, Carrie. (2003) "Local Level Challenges to Post-Conflict Peacebuilding," *International Peacekeeping* 10(3), pp. 25–43.

Murithi, Tim. (2009). "The African Union's foray into peacekeeping: Lessons from the hybrid mission in Darfur," *Journal of Peace, Conflict and Development*, Issue 14, pp. 1–19.

Nkrumah, Kwame. (1976) *Neo-Colonialism: The Last Stage of Imperialism*. New York: International Publishers.

Pankhurst, Donna. (2010) "The 'Sex War' and Other Wars: Towards a Feminist Approach to Peace Building," *Development in Practice* 13(2–3), pp. 154–177.

Paris, Roland. (2002) "International Peacebuilding and the 'Mission Civilisatrice,'" *Review of International Studies* 28(4), pp. 637–656.

———. (2010) "Saving Liberal Peacebuilding," *Review of International Studies* 36(2), pp. 337–365.

Puechguirbal, Nadine. (2012) "The Cost of Ignoring Gender in Conflict and Post-Conflict Situations: A Feminist Perspective," *Amsterdam Law Forum* 4(1), pp. 4–19.

Richmond, Oliver P. (2009) "The Romanticisation of the Local: Welfare, Culture and Peacebuilding," *The International Spectator* 44(1), pp. 149–169.

Richmond, Oliver P. (2003) "Introduction: NGOs, Peace and Human Security," *International Peacekeeping*, 10(1), pp. 2–11.

Richmond, P. Oliver, and Mac Ginty, Roger. (2015) "Where Now for the Critique of the Liberal Peace?" *Cooperation and Conflict* 50(2), pp. 171–189.

Sabaratnam, Meera. (2017) *Decolonising Intervention: International Statebuilding in Mozambique*. London: Rowman & Littlefield International.

Schoeman, Maxi. E. (2002) "Imagining a Community: The African Union as an Emerging Security Community," *Strategic Review for Southern Africa*, *24*(1), pp. 1–26.

Tieku, Thomas Kwasi. (2012) "A Pan-African View of a New Agenda for Peace," *International Journal* 67(2), pp. 373–389.

Touray, Omar A. (2005) "The Common African Defence and Security Policy," *African Affairs* 104(417), pp. 635–656.

United Nations (UN). (1992). *An Agenda for Peace: Preventive Diplomacy, Peacemaking and Peacekeeping.* New York: United Nations.

United Nations (UN). (2017, July 19) Secretary-General, at Security Council, stresses shared interest of United Nations, African Union, in stronger conflict prevention mechanisms. Security Council, 8006th meeting.

Weiss, Thomas G. (2004) "The Sunset of Humanitarian Intervention? The Responsibility to Protect in a Unipolar Era," *Security Dialogue*, 35(2), pp. 135–153.

Welsh, Jennifer M. (2013) "Norm Contestation and the Responsibility to Protect," *Global Responsibility to Protect*, 5(4), pp. 365–396.

Williams, Andrew P., and Berhanu Mengistu. B., (2015) "An Exploration of the Limitations of Bureaucratic Organizations in Implementing Contemporary Peacebuilding," *Cooperation and Conflict*, *50*(1), pp. 3–28.

World Bank. (2002) *World Development Report 2002: Building Institutions for Markets.* Washington, DC: Oxford University Press.

Yamashita, Hikaru. (2012) "Peacekeeping Cooperation between the United Nations and Regional Organizations," *Review of International Studies*, pp. 165–186.

Zanotti, Laura. (2011) *Governing Disorder: UN Peace Operations, International Security, and Democratization in the Post-Cold War Era.* University Park, PA: Penn State University Press.

4 The ethics of research and fieldwork in conflict-affected African settings

Ismail Rashid[1]

Introduction

Scholars cannot responsibly engage in research relating to violence, conflict and peacebuilding in Africa without serious ethical considerations. This is even more imperative with regard to studies in conflict-affected settings (Porter, Robinson, Smyth and Osaghae 2005). Working in such settings can affect the security and welfare of researchers, collaborators and subjects, the effectiveness of their methodologies, the quality of their data and the integrity of their analysis and findings (Wood 2006; Ford, Mills, Zachariah and Upshur 2009). It is difficult to know precisely how researchers would respond to ethical challenges in conflict-affected settings, but some may be inclined to take short cuts and behave unscrupulously. Individual ethical disposition, one's sense of morality and behavior are conditioned by multiple factors, and virtually every human society has principles, rules and practices relating to right and wrong behavior. Research ethics are anchored on principles such as respect, beneficence, justice and integrity, which many societies, cultures and religions accept as indicative or supportive of good moral disposition and actions.

These principles have become embedded in laws and regulations governing research in many parts of the world, including in some African contexts (Fujii 2012; Ndebele, Mwaluko, Kruger, Ouwe Missi Oukem-Boyer and Zimba 2014; Carpenter 2018). Researchers are required to get approval from institutional ethics review bodies before commencing any field research. The ethics review process usually scrutinize all aspects of the research, including topics, questions, sites, participants, methodology and data collection strategies to ensure privacy, protection, reliability, integrity and minimization of harm. Whether engaged in clinical trials, conducting focus group discussions, interviewing people or doing participant-observations, researchers and their assistants are required to act in morally acceptable and legally compliant ways.

Yet, even the most diligent researchers and robust ethics review processes cannot anticipate all of the exigencies that may be present in the field, especially in conflict-affected contexts. There is no comprehensive list of advice that could be offered to researchers in terrains of complexity, unpredictability and higher levels of risks than normal. Some exigencies may demand reflexive, improvised or creative responses on the part of researchers (Cohen and Arieli 2011). These responses can sometimes place researchers in an ethical quandary and they can resort to relativistic and "situational ethics." Situational and relativistic ethics, an area of much debate, can be a slippery slope and may result in the very harm that research ethics aim to prevent.

While situational responses are important in thinking about research ethics, considerations of ethics however need to be much broader in studying conflict and peacebuilding in Africa. Scholars like Carpenter argue for "virtue ethics," which entail continuous reflections in all procedural aspects of the research (2018). While I agree, I will suggest like Fujii that the application of virtue ethics should not be limited to procedural aspects, they should be more comprehensively applied, encompassing all aspects of research projects, from conception to completion (Fujii 2012). As such, researchers need to think about their project in relationship to the epistemic struggles over the production of knowledge about Africa and Africans and the use of the continent and its peoples as a research laboratory and a site of extraction. Are they the collectors of empirical data for others to theorize or the generators of new ideas and theories that contribute to the store of human knowledge that is beneficial to Africa? Ethical considerations should also infuse their understanding and assessment of their "conflict-affected" field, especially how it developed, how it is constituted and its specific characteristics.

Once they are in the field, researchers need to be attuned to the relationships, emotions and interactions that will shape their ethical responses. Creative and reflexive actions may sometimes be unavoidable, but ethical consciousness, active planning and careful preparations are vital. All ethics review processes have their contradictions; however, they should not be ignored or viewed as simply a bureaucratic inconvenience. They should be utilized as part of the education and reinforcement of the researchers' understanding of research ethics and risk assessment and mitigation. In the field, researchers should use informed consent and conduct research activities in active and dynamic ways so as to demonstrate integrity as well as build the kinds of trust, reciprocity and mutuality that gives meaning to research ethics and the research endeavor in general.

The main focus in this chapter is on the social sciences rather than the biomedical sciences. Documented cases of egregious experimentation and clinical trials on people in Nigeria, Zimbabwe and South Africa, some with fatal outcomes, have spurred greater awareness and action on the part of biomedical researchers to ensure that such activities in Africa are in line with global ethical principles and norms (Kruger, Ndebele and Horn, 2004; Falusi, Olopade and Olopade 2007; Ndebele, Mwaluko, Kruger, Ouwe Missi Oukem-Boyer and Zimba 2014). While both African and non-African researchers are expected to conform to the highest ethical standards in their research, non-Africans can sometimes utilize the power and resources of their wealthier countries, universities and funding agencies to respond to risks and ethical challenges in the field in ways that Africans cannot (Porter, Robinson, Smyth, Schnabel and Osaghae 2005; Thomson, Ansoms and Murison 2012). This chapter is structured into three sections; the first reflects on the epistemological challenge posed by research on Africa and Africans; the second discusses the ethical risks and complexities of researching in conflict-affected environments; and the third considers ways that researchers could use informed consent and their research activities to build trust, transparency and reciprocity.

Conflict Africa: laboratory Africa. The field as an episteme

If researchers take seriously the principles that are at the heart of research ethics, then any study of conflict, violence and peacebuilding in Africa should instinctively and reflexively provoke interrogation. There has been much debate over the production of knowledge, power and hierarchy, and how Africa and Africans have been marginalized and disempowered (Medie and Kang 2018; Obeng-Odoom 2019). Similarly, there is much critique of the relationship between the North and South, especially when it comes to the division of labor regarding research and its various uses (African assistants and scholars do the challenging fieldwork and data collection; scholars from the North interpret and theorize this data) (Mamdani 2011). It is not surprising that many African scholars have called for the decolonization of, epistemic rupture from and the recalibrations of these types of knowledge production and utilization (Ndlovu-Gatsheni 2020; Musila 2017). The study of violence, conflict and peacebuilding is squarely implicated in this broad debate about the nature and ethics of knowledge production. It is a field that sits uneasily at the intersection of Africa's recent realities, research into these realities and the generation of data, analysis, theories to understand these realities.

Conflict, violence and peacebuilding, subjects whose contours are evolving and constantly being redefined, did not generate the realities of conflicts in Africa. They have merely responded and derived much intellectual oxygen from the research of these realities. The contemporary recent realities in Africa are indisputable. The postcolonial period, particularly the decades following the collapse of the post-World War II bipolar world, has been particularly turbulent and lethal for Africans. Newer, deadlier forms of intrastate and interstate violence combined with older forms of conflict to wreak extensive havoc in all regions of Africa. In the 1990s and early 2000s, coups d'etat and civil wars blighted Sierra Leone, Liberia, Cote D'Ivoire and Guinea Bissau, while secessionist violence bogged down the energies of central governments in Senegal, Nigeria and Mali in west Africa (Adebajo and Rashid 2004). In central Africa, Rwanda and Burundi experienced civil wars and the horrors of genocide, while the Democratic Republic of Congo had a series of complex civil and regional wars (Khadiagala 2006).

Other African regions also experienced significant conflicts. In north Africa, the Algerian government fought a decade-long civil war against various Islamist groups to keep them out of power. In the horn and east of Africa, protracted wars eventually led to the independence of Eritrea from Ethiopia in 1993 and South Sudan from the Sudan in 2011. These outcomes did not stop Ethiopia and Eritrea from fighting between 1998 and 2000, end the conflicts in Darfur and other regions in the Sudan or prevent South Sudan from subsequently having its own civil war. Somalia's ever shifting civil war is now in its third decade, even after interventions by UN, Ethiopian and African Union forces. In southern Africa, where liberation from colonialism and white supremacist regimes were delayed, considerable violence accompanied the dismantling of Apartheid in South Africa, and the rebuilding and democratization of Zimbabwe and Mozambique after their liberation wars.

While many of the intense and violent conflicts of the 1990s and 2000s have ebbed or been resolved across the continent, a few have morphed and have flared up again, and newer ones have emerged. The violence in Nigeria's delta region has been succeeded by the insurgency of Boko Haram in the northeast and many other smaller conflicts across the country. The recrudescence of a separatist uprising in northern Mali, which started in 2012, has now become part of a trans-Sahelian conflict complex involving armed secessionists, jihadists, ethnic militias and transnational troops. In north Africa, Libya has become the most prominent theater of intrastate violence following the "Arab Spring," the NATO bombing and assassination of Gaddafi, and state fragmentation from 2014 onwards (Campbell 2013).

As indicated earlier, these extensive post-colonial and post-cold war conflicts in Africa have powered a research and publication boom amongst African and non-African scholars, working independently or in collaboration. In the introductory chapter to one such collaborative study, *Researching Conflict in Africa: Insights and Experience* (2005), Eghosa Osaghae and Gillian Robinson assert:

> Africa has the uncanny reputation of being the world's leading theatre of conflict, war, poverty, disease and instability. Therefore it is not surprising that scholars of ethnicity and conflict management regard it as a major laboratory for experimentation and theory building. Notwithstanding the exaggerations and oversimplifications that sometimes attend the claims and findings, including the tendency to lump all states in the continent as suffering from the tribalism disease, Africa generally has not disappointed and, in a manner of speaking, has lived up to its billing.
> (2005, p. 1)

Researching conflict in Africa is an important text in many ways, especially in critiquing earlier analyses of conflict and politics, drawing attention to the international dimensions of local conflicts and foregrounding the field experiences of different researchers in a number of African countries. Yet it is striking that Osaghae and Robinson would invoke the notions of Africa as conflict-prone and as a "laboratory" without significantly interrogating them.

As noted earlier, the realities of conflict in reshaping and mediating politics and the lives of millions in Africa in the last three decades are undeniable. What is at issue here are the different discourses that are invoked to problematize, signify, interpret and understand these realities. Conflict-prone Africa is an important signifier in the discursive apparatuses of European and Western racializing and colonizing perceptions of Africans. It provided justification for European conquest, domination and exploitation of Africa in the late 19th century, and it still continues to undergird Europe's unequal relations with the continent and its peoples. The *pax Britannica, mission civilisatrice* and various other European imperial projects in Africa conceived of Africa and Africans not only as objects to be dominated, exploited and civilized, but also as subjects of all kinds of studies, experimentation and theorization. The British colonial administrator, William Malcolm Hailey, aptly captures the disposition in his *African Survey*, when he stated, "Africa presents itself as a living laboratory, in which the reward of study is to be not merely the satisfaction of an intellectual impulse, but an effective addition to the welfare of a people" (1938).

Europeans scholars, scientists, administrators or even dilettantes not only collected empirical information and all kinds of materials but also conducted extensive research on all aspects of Africa and African life (Tilley 2011). Colonial biomedical and social sciences, old and new, operated in this African "laboratory," sometimes with grave consequences for African health, lives and dignity. The knowledge from "laboratory Africa" fed into various technologies of imperial exploitation and domination, and it also produced pernicious cartographies of power and racial taxonomies that continue to structure global geopolitics. Generations of African scholars have critiqued these processes of knowledge production and the power hierarchies they create and sustain (Mudimbe 1988; Ogot 2009). These critiques have occasionally forced retreats and recalibration of these processes, but the hierarchies and taxonomies merely mutate and reappear in new guises because of the power and resources that continue to be invested in them.

For all of the reasons mentioned earlier, even before researchers venture into the field, they need to reflect on the ethical issues arising from their choice of topics and how it relates to the historic struggle around epistemic dominance, resistance and the quest for freedom and justice in Africa. Researchers studying conflict and peacebuilding are at the heart rather than the periphery of this epistemic struggle in the 21st century. In the last three decades, considerable resources and scholarly energies have been expended and publications have proliferated on conflict and peacebuilding in Africa. Partly as a result of the neoliberal marketization of education and research from the 1990s onwards, this research has been conducted under the auspices of universities and institutes as well as through various international organizations, governments, nongovernmental organizations, think tanks, foundations and multinational corporations (Ford, Mills, Zachariah and Upshur 2009). The proliferation of conflict and peacebuilding research funding and studies places conscientious African researchers in an ethical bind. On the one hand, it seems that such research provides the empirical evidence that reinforces the problematic discursive framings of Africa and its peoples. On the other hand, there is a compelling need to understanding the different forms of violence, their continuing heavy toll on Africa and Africans and how they are constraining or opening up new possibilities of freedom, justice and development.

Studying violence should not be conceived as a "simple endeavor" or an academic exercise to foster the career of a scholar. As Nordstrom and Robben contend, it is an endeavor "fraught with assumptions, presuppositions, and contradictions" and "an intricately layered phenomenon" (1995). So, researchers have to ask themselves about why they are gravitating towards the field and researching particular topics? Is it out of academic curiosity, the availability of resources or their potential emancipatory value? Have they sufficiently interrogated the ideas, theories and methods that are deeply embedded in such research projects and are they prepared to have them challenged, revised or even replaced in the field? Regardless of their answers, researchers have to realize that history and context matters, and that an awareness of that is an integral part of their ethical accounting. They cannot go to the field without understanding the past of that "field" and how it continues to be molded by all kinds of forces, local, national and international. Researchers cannot uncritically accept the narratives of Africa in many of the existing scholarship, and they have to reflect on the extent to which their research may be reinforcing or challenging the discourses that disempower Africa and Africans.

Conflict zones, unpredictable terrain: the field as a place and peoples

A complex landscape

The "state" of the actual research field can have serious ethical implications for researchers. The timing, complexities and dynamism of conflicts in Africa mean that there is no archetypal research field, and that research terrains fall on different parts of a grid. The most extreme and most dangerous terrains are those in which there is active and widespread violence and insecurity. Except for very limited areas, local, national or international authorities or forces are unable to provide significant or sustained protection for people in these countries. While many African countries were in this state in the 1990s and early 2000s, few if any (perhaps Libya and Somalia) fits this mold at the time of writing. While some researchers, out of a sense of bravado or being on the cutting-edge of scholarship, may want to see the dynamics of "conflict" or "laboratory" Africa in real time, I doubt whether any conscientious ethics review committee would approve such a research.

In the middle of this grid are those research terrains where the conflicts are regionalized or localized, political centers remain strong, security forces are operational and large parts of the society are relatively stable and functional. These regionalized or localized conflicts can be protracted, intermittent or sporadic. Nigeria, Mozambique, Mali and many African countries have research fields which fall within this part of the grid (see Babatunde 2019; Duruek 2020). Even with the risks involved, many research committees would approve research in such terrains because they are usually conducted on the edges rather than in the crucibles of these conflicts. Finally, at the far end of the grid are research terrains which are in post-conflict countries and where violence has ended or receded significantly. The state and society are generally stable and functional enough to support field research. Uganda, Sierra Leone, Liberia, Rwanda, Angola and the majority of African countries fall within this part of the grid. For researchers, while it is important to recognize the complexities and variations of the different research "fields," it is also important to push back against totalizing assumptions of the state of conflicts in the continent. The context of complex wars involving known and emerging dynamics makes it particularly difficult to systematize, model or classify conflicts.

Yet, it is indisputable that all conflicts disrupt societies, and the most violent ones leave destruction, death and trauma in their wake. They also wreak havoc on infrastructure, economies and the social rhythms by which people sustain themselves. Contestations of power are usually palpable and visible in the presence of various armed groups and security agencies. Many people living within or near conflict-affected environments have heightened vulnerability. Even in post-conflict societies, where conflicts are receding or have long receded, have receded or are receding, the aftereffects of violence, trauma and insecurity are usually visible in the landscape, in peoples' conditions, how they act and how they interact with others. These landscapes are usually highly emotionally charged and outsiders are viewed with suspicion and distrust (Cohen and Arieli 2011). Grievances, distrust, social tensions and hostilities are either apparent or lurk beneath the surface.

Whether the researcher is an insider or outsider, research fields across the world can be dangerous, risky and full of obstacles. Nancy Howell's pioneering report, *Surviving Fieldwork*, provides comprehensive insight into some of the political, health and general hazards of anthropological research in different parts of the word, including those experiencing political and social upheavals. Researchers suffered from disease, mental exhaustion or deterioration, road accidents, substance or drug abuse, physical violence, kidnapping, arrests, detention, reprisals and death for researchers and participants (Howell 1990; See also Fujii 2012, p. 717). Even though produced three decades ago, the report remains a valuable primer for anthropologist and other social scientists undertaking field research. Ayantayo (2014) provides a useful toolkit of 23 ethical codes and 19 safety practices for African researchers working on peace and conflict.

Knowing the state of the "field" (pun intended) is a crucial starting point for all researchers, whether they are nationals or foreigners, or insiders or outsiders. Being an insider or outsider comes with their opportunities, baggage and dangers into the field. There is much debate about insider-outsider dynamics elsewhere (Toy-Cronin 2018). Being citizens or residents of a country might mean that researchers have some base knowledge of the research field, but in many cases this is hardly sufficient; distant or even adjacent regions may just as well be foreign terrains to them. All researchers should endeavor to have meaningful knowledge of the environmental characteristics, peoples, histories, cultures, social dynamics, politics and ongoing state of affairs as a starting point. Even researchers who have done previous research in a field known should be aware that all landscapes change, more so if they have been affected by conflict. Researchers should undertake exploratory field visits and small pilot studies to assess feasibility, timing and risks associated with a research study (Sluka 1995, p. 282). Such visits and pilot studies are also important for researchers and their assistants to build relationships and to develop contingent and emergency plans to deal with anticipated risks.

Once researchers are in the field, they must be conscious (and some are) that they are in a milieu in which the entanglements of environment, production, histories, cultures and politics have actively shaped people and their relationships. These relationships assume different characteristics in different African countries. In different conflict-affected African settings, researchers have to be conscious and reflective about how different people, relationships and interactions generate various kinds of ethical jeopardy or risks. There is another dimension of active planning that is worth emphasizing, it is an assessment of a real risk for a researcher to "securitize" a research field and research subjects. At the core of these ethical jeopardy and risks are perceptions, emotions, differential power and calculations, which may cause people, including researchers, to act in immoral and unethical ways. It is impossible to identify and discuss all the groups or issues that may arise; I want to focus on four broad groups – vulnerable people, institutional facilitators, security agencies and armed groups and teaching assistants.

Vulnerable people

There are social scientists who view the notion of vulnerability as problematic because it was initially introduced for biomedical research, tends to be "fixed and unmovable," "vague" and casts researchers as potentially "dangerous." Consequently, van den Hoonard argues, "we are better off disowning the term" and should "send it to the dustbin of unworkable and imaginary concepts" (2018, p. 305). While there is some merit in the critique of the application of term, it would be unethical not to consider vulnerability in conflict-affected situations. Rather than perceiving vulnerability as fixed, researchers should approach it as determined by status as well as being situational and relational. Vulnerability may be determined in conflict-affected contexts by age, gender, ethnic, religious and national identity, ability, sexuality and war impact. It applies to people who are elderly, women, youth, children, physically handicapped, prisoners and minority groups, who are susceptible to be, or have been victimized, traumatized and exploited. Such vulnerable people are usually the subjects of and participants in many conflict and peacebuilding studies, and they could rightfully be resentful or cynical towards researchers (Babatunde 2019). These groups usually have limited resources and abilities to protect themselves from those who carry weapons, have financial or other material resources or wield power. Research ethics are designed primarily to protect them the most.

Vulnerable individuals or groups may not necessarily view researchers as people that they need protection from, except when their unethical actions are egregious and become public. In many instances, however, they do perceive researchers as wielders of resources and power, depending on how researchers enter and present themselves in conflict-affected settings and the subject of their research. They may rightly or wrongly surmise that researchers have strong connections to local, national or international agencies (Babatunde 2019). This may lead to a "misinterpretation," if not exaggeration, of the amount of power and authority that researchers actually wield. Such perceptions and misinterpretation can put researchers in an ethical conundrum as to how they are expected to act in that setting.

Apart from the ethical conundrum that comes with perceptions of power, researchers may also have to deal with the emotions, interests and calculations of vulnerable individuals and groups. Within and outside the research process, the situation of vulnerable and marginalized people can generate a whole range of emotions from researchers, including intense sadness, pity, empathy (Wood 2006) or even disdain. As dynamic and active agents, vulnerable individuals and groups can also leverage these emotions in ways that can be challenging for the researcher and the research project. Vulnerable and marginalized people may rightly develop strong stakes in the narratives being generated about them, viewing research projects as avenues to convey the realities of their situation and to advocate for their needs. How does the researcher deal with these humanely and respectfully deal with these emotions and interests without becoming overtly paternalistic, cynical or even disrespectful?

Powerful institutional actors

Institutional facilitators in a conflict-affected environment may also pose ethical challenges for researchers, their assistants and collaborators. They are mainly the members of governments and their related agencies, various local, national, international governmental and nongovernmental organizations and businesses who create access, connections and provide various kinds of resources for researchers. People, who have emerged in conflict-affected environments as informal brokers, could also be part of these groups. It is a complex group of people who have different stakes in the conflict and wield different kinds of power and authority that can be used beneficently or maliciously.

Depending on the subject of the research, members of governments, organizations and businesses may have a strong stake in the data being collected and the findings, and can seek to influence it in one way or another. They may act as gatekeepers and try to control the scope of the researchers' contacts and interactions as well as influence their narratives. Researchers are sometimes dependent on some of these facilitators, their organizations and networks for their personal safety in conflict-affected settings (Porter, Robinson, Smyth, Schnabel and Osaghae 2005; Thomson, Ansoms and Murison 2012). They can therefore be seen as mediators of the research process. They have to trust them, but association with them can create problems in the field. How should the researcher manage these relationships without being complicit in injustice or being detracted by awe, suspicion or indignation?

Another group with considerable potential to impact researchers' welfare and actions on the group are national or transnational security agencies, militias, rebel movements, vigilantes, gangs and other armed groups. Even when women are members of these groups, the environment around them is usually hyper-masculinized. They can be facilitators and protectors for researcher and participants or obstructers and sources of harm for them (Duruek 2020). Since they are sometimes directly involved in perpetrating, containing or ending violence or ensuring security, they therefore tend to be very secretive or hyper-vigilant about information collection activities, even scholarly ones.

Security agencies and other armed groups are particularly averse to research that can reveal crimes, atrocities, human rights violations and other forms of illegal and unethical behavior, though some of these groups (or individuals within them) have no such aversion. For some researchers, interactions with some of these groups may be tinged with intense anxiety, fear and perhaps anger (Wood 2006). How do researchers protect themselves and participants from the negative emotions, dangers or mistrust generated by interactions with such groups in their research?

Research assistants

Researchers' perceptions of and relationship with research assistants in the field are crucial. Assistants are students or researchers or people from the general population contracted to help with logistics, translation and transcription and data collection and analysis. Depending on the tasks, researchers and their assistants could be in frequent and close contact in the field and therefore be exposed to similar dangers and challenges. The range of emotions that can arise in the relationship between researchers and their assistants cannot be easily catalogued. Nonetheless, it is a frequently hierarchical relationship, in which the power and resource balance is in favor of researchers. Assistants have been and can be subjected to exploitation, abuse and poor treatment from researchers as well as participants in a research project.

Researchers must endeavor to protect their assistants from exploitation, discrimination, harassment and harm during their employment. Even if these assistants are undergraduate or undergraduate students under the supervision of the researcher, they must be fairly compensated, especially if their employment is included in the research funding. Researchers have to make sure that assistants are sufficiently trained and made aware of their tasks and responsibilities in the field and in the overall research project. They should also ensure that the contribution of research assistants is duly acknowledged in the research outputs.

Navigating the field

Reflecting on the personal

Researchers' responses to emotional and ethical challenges in the field would most certainly lie in the interaction between the circumstantial, the structural and the personal. Regarding the personal, just as there is no archetypal field, there is no archetypal African researcher. Researchers are usually a plenum of identities, intersected by gender, race, religion, nationality, ethnicity, sexual orientation and other characteristics. They are also practitioners of particular disciplines and crafts. The different personal and disciplinary identities, separately or in combination, will certainly mediate their ethical disposition, choices and actions or even the kind of data they acquire (Yacob-Haliso 2018). If they are applying virtue ethics:

> [t]he qualitative researcher cannot be an objective researcher or an objective bystander, collecting data in a personally disinterested fashion; they should become immerse in their work and be aware of the emotions it evokes and the presence and impact of their personal values.
>
> (Carpenter 2018, p. 36)

Though the identities of the researcher and circumstances in the field may be what drive their responses in field, it is important they remain cognizant of how they can utilize structures and activities that are already constitutive of their research project and ethical review processes. Researchers have to use informed consent and their research activities in the field to build trust, reciprocal relations and mutuality in order to minimize and contain incidents that may put them in ethical jeopardy. They have to reflect on and pay particular attention to areas like compensation and transparency where issues of respect, justice and beneficence come up in stark relief.

Cultivating a mindset of accountability

Researchers in conflict-affected environments have to cultivate a mindset of ethical accountability. For some social researchers, ethic reviews committees are an affront to academic freedom and an attempt to impose Western values and "principlism" in non-Western societies, and largely derivative from biomedical sciences (Israel 2018). In institutions which have either robust or even nondescript ethical committees, they can be Kafkaesque entities with laborious bureaucratic requirements which unnecessarily hold up research projects. Ethics review committees, where they exist in Africa, vary greatly in quality and efficiency. The effectiveness of these committees is usually hampered by lack of membership diversity, inadequate training of members, scarcity of resources and absence of national regulations (Nyika et al. 2009; Sleem et al. 2010). The situation in many African institutions can be far from ideal in terms of ethics review, and some of the principled objections about ethical "imperialism" are worth further debating.

However, this should not obviate the necessity of researchers seeking ethical oversight, especially given the history of research malfeasance in the continent (Mutenherwa and Wassenaar 2014). Researchers in Africa should start with ensuring their proposals undergo scrutiny, vetting and approval from institutional ethics review bodies, where they exist. Many African states and universities now have such bodies (Kruger, Ndebele and Horn 2014). The review process is a concrete reminder that how researchers behave in the field is not simply a matter of personal propriety, it also is usually regulated by laws, policies and procedures established by governments, professional organizations and academic institutions. So, going through ethics review process is an issue of academic accountability and compliance with the law, even if the efficacy of the law and procedures are debatable.

Ethics review bodies and processes are gatekeeping processes, and unfortunately can sometimes become adversarial for African researchers. Their role is not to set research agenda, topics or constrain the academic freedom of researchers. Yet, they are important steps in scrutinizing the appropriateness of the research, the qualification of researchers and challenges that researchers may encounter in the field and of which they may be unaware. Where they are robust (and African scholars must actively ensure that they are), they will be attentive to efficacy of methodologies, timing, research activities, research locations, selection of participants; compensation and exchanges of money; security, privacy and confidentiality; storage and protection of data; and levels of risks in the field. In making sure that researchers are only compliant with regulations, ethics review bodies can help researchers anticipate, minimize and take adequate precautionary measures. This can help create layers of protection for all involved in a research project (Levasseur 2000). More than just ticking off boxes, researchers must see the ethical review process as integral to the conception, design and execution of the research as well as cultivating a mindset of accountability.

Building consent in field

One of the areas where that mindset of ethical accountability is operational in the field is the process of acquiring informed consent. Consent from participants in a study is expected to be given in a transparent, verifiable fashion, either as a signature or thumbprint in a prepared form or recording. This should be done after researchers or their assistants have explained their study, the nature and extent of participation required and the rules of engagement and disengagement. It's a contract which ensures that the parties in the study are participating voluntarily, freely and with some understanding of what is involved in the research. It should also cover recordings, images and uses of data. But as Wasuna, Tegli and Ndebele underline in their reflections on ethics of biomedical research in Africa, it

> means more than simply obtaining the signature of the research participant. It is a process that involves conveying accurate and relevant information about the study, its purpose, potential benefits, known risks, alternatives and procedures, in a language that the participant best understands.
>
> (2014, p. 58)

As a process, ensuring informed consent entails a number of considerations that are particularly important in war-affected environments. The environment or setting within which it is acquired should be noncoercive or one inappropriately incentivized. While some settings might be convenient, researchers should be particularly sensitive to how power holders – government officials, chiefs, security officials, corporate bosses, community leaders and so on – are "helping" to facilitate consent. As much as possible, researchers have to ensure that participants are mature, able and/or have permission and support (in the case of minors) to participate willingly. The process of securing informed consent needs to be culturally sensitive and locally appropriate, and acquired in the languages spoken by people in the various communities that the research is being done. That the participants or population is "illiterate" or "under-educated" is not a valid excuse; it means that researchers should have more patience and diligence in explaining the research and research activities.

Researchers have to be as transparent as possible, although there may be some instances where transparency could clash with the integrity of the research or other important ethical principles. Opaqueness may be required. Anonymity and confidentiality may also be crucial for the safety of participants and researchers. There is much scholarly work on how to ensure anonymity in the data collection and analysis. However, researchers need to be very careful about information that will endanger, compromise or render them liable to prosecution. Some researchers have gone to great lengths, enduring threats and arrests to protect anonymity and confidentiality, especially against autocratic regimes, security agencies and even rebel forces (Thomson, Ansoms and Murison 2012). African researchers, unlike Western ones, tend to have less ability to push back against these governments, security agencies or forces (Ibid). While transparency and discretion may not always prevent misunderstanding, misinterpretations, suspicions, they can help set the expectations from your side, and can inoculate you and your assistants against potential risks.

Acquiring consent and being transparent should not be one-off acts, but an engaged and sustained process of education and socialization of participants about the research. It involves patiently listening to participants' questions and providing sufficient answers. It entails not creating unrealistic expectations of research findings and what a researcher can do to influence local, national and international policies or actions. Researchers should be prepared to do this often, in individual interviews, administering surveys, focus group discussions and collection of life histories. In conflict-affected settings, the research topic or agenda may not be a priority or congruent with the needs and interests of the participants; researchers have to accept this reality. In short, informed consent should not simply be seen as a legal and bureaucratic requirement, but an opportunity to establish relationships based on trust and mutuality (Ibid).

Relationships of trust and mutuality are particularly important in qualitative research activities where data acquisition can be interactive and dynamic, namely interviews, collection of life histories, participant observation, focus group discussions and validation sessions. Generally, it is crucial that participants in the research are treated with dignity, respect and fairness. As mentioned earlier, researchers and their assistants should understand the social dynamics and cultural rules of interactions, some of which may be unpredictable, fluid and under contestation. They should incorporate these dynamics and rules into methods and activities whenever possible. There are fields where interactions between people of different genders are considered inappropriate, and this should be taken into consideration in interviews, surveys, focus group discussions and other research activities.

Researchers and their assistants have to be careful about their questions and how to facilitate discussions. They should know the kinds of questions that may be taboo, and are likely to reignite latent tensions, inflame hostilities or affect peacebuilding efforts. Even when participants seek to make research activities platforms for competing narratives, researchers and their assistants should try not to be drawn into local, national or international power struggles. Apart from potentially having negative outcomes, these struggles sometimes take considerable skill and energy to navigate and disengage from.

Where there are dangers of traumatization or re-traumatization of individuals, groups and communities, researchers need to ensure in advance that there will be the requisite psychosocial support or counseling.

Compensation and fairness

Compensation and gift-giving can also be a fraught terrain for researchers, and may affect the disposition, actions and contributions of participants in research activities and the quality of the information they provide. This is particularly salient in conflict-affected environments where resources could be scarce, and research activities are sometimes impinging on participant's time and ability to seek resources they need. As some scholars have pointed out, research can be a process of unequal exchange and extraction that is more beneficial to researchers Even if the research is relevant to the interests and welfare of participants, its benefits may not be immediate or direct. (Wolf 1996, Wood 2006).

Compensating participants in research is becoming less of a contentious issue. Partly as a result of the marketization of scholarly research in the last few decades and serious considerations of issues of justice, many ethics committee approve some form of compensation or gift-giving. Participants do invest time, labor and sometimes their own personal resources in research activities. In the biomedical sciences, where there has been a history of sometimes compensating participants in research and clinical trials, the debate has been more about different models of compensation – market, wage, reimbursement or appreciation (Dickert and Grady 1999). Whichever model a researcher decides on needs to be sensitive to the cultural and socio-economic context, particularly in conflict-affected areas (Mite and Ndebele 2014).

Conclusion

Researchers working in conflict-affected environments of Africa will undoubtedly face higher risks and ethical challenges. How researchers behave in conflict-affected environment is contingent upon a number of personal, structural and situational factors. Hopefully, they would be ethically aware and have sufficiently prepared for these challenges. Awareness and preparation should start at the conception of the research project with the understanding that the production of knowledge is never a neutral endeavor, and that it sometimes mirrors the struggles over meanings, power and resources in the field. Therefore, researchers have to be conscious of the relationships and the emotions surrounding their research, even as they are alert to situations that may be ambiguous, compromising or dangerous. In such situations, facilitators, relationships and networks are crucial in producing or mitigating harm. Informed consent and research activities should be utilized as opportunities to build trust, reciprocity and respect. Ultimately, researchers need to be sensible and adaptive in their responses, and not assume that their preparations, experience and instincts will be sufficient.

Note

1 Support for research on this chapter was provided by the Vassar College Elinor Nims Brink Fund.

References

Adebajo, Adekeye and Rashid, Ismail. (2004) *West African Security Challenges: Building Peace in a Trouble Region*. Boulder, CO: Lynne Rienner.

Ayantayo, Jacob Kehinde. (2014) "Ethical Issues in Peace and Conflict Research," in Albert, Isaac Olawale, Olasehinde-Williams, Olabisi, and Aremu, Oyesoji (Eds.), *Research Methods and Peace and Conflict Studies in Africa*. Ibadan: University of Ibadan Press, pp. 269–290.

Babatunde, Abosede Omowumi. (2019) "Renegotiating Gender during Fieldwork in Conflict-Affected Settings: Overcoming Some Challenges," *Kujenga Amani*, March 14. https://kujenga-amani.ssrc.org/2019/03/14/renegotiating-gender-during-fieldwork-in-conflict-affected-settings-overcoming-some-challenges/ (Accessed June 3, 2020).

Campbell, Horace. (2013) *Global Nato and the Catastrophic Failure in Libya: Lessons for the Forging of African Unity*. New York: Monthly Review Press.

Carpenter, David. (2018) "Ethics, Reflexivity and Virtue," in Iphofen, Ron and Tillich, Martin (Eds.), *The Sage Handbook of Qualitative Research Ethics*. London: Sage, pp. 35–50.

Cohen, Nissan, and Tamar Arieli. (2011). "Field Research in Conflict Environments: Methodological Challenges and Snowball Sampling," *Journal of Peace Research*, 48(4), pp. 423–435.

Dickert, N., and Grady, C. (1999) "What Is the Price of a Research Subject: Approaches to Payment for Research Participation," *The New England Journal of Medicine*, 34(3) pp. 198–203.

Durueke, Oyinyechukwu Onyido. (2020) "Survivors of Sexual Violence: Navigating Post-Conflict Environments in Nigeria's Niger Delta," *Kugenja Amani*, February27,2020. https://kujenga-amani.ssrc.org/2020/02/27/survivors-of-sexual-violence-navigating-post-conflict-environments-in-nigerias-niger-delta/ (Accessed June 3, 2020).

Falusi, A. G., Olopade, O. I., and Olopade, C. O. (2007) "Establishment of a Standing Ethics/Institutional Review Board in a Nigerian University: A Blueprint for Developing Countries," *Journal of Empirical Research on Human Research Ethics*, 2(1), pp. 21–30.

Ford, Nathan, Edward J. Mills, Ronny Zachariah, and Ross Upshur. (2009) "Ethics of Conducting Research in Conflict Settings," *Conflict and Health*, 3, p. 7.

Fujii, Lee Ann. (2012) "Research Ethics 101: Dilemmas and Responsibilities," *PS: Political Science & Politics*, 45(4), 717–723.

Hailey, William Malcolm. (1938) *An African Survey: A Study of Problems Arising in Africa South of the Sahara*. Oxford: Oxford University Press.

Howell, Nancy. (1990) *Surviving Fieldwork*. Washington, DC: American Anthropological Association.

Israel, M. (2018) "Ethical Imperialism? Exporting Research Ethics to the Global South," in Iphofen, Ron and Tillich, Martin (Eds.), *The Sage Handbook of Qualitative Research Ethics*. London: Sage Publications Ltd., pp. 89–100.

Khadiagala, Gilbert. (2006) *Security Dynamics in Africa's Great Lakes Region*. Boulder, CO and London: Lynne Rienner.

Kruger, Mariana, Ndebele, Paul and Horn, Lyn (Eds.) (2014) *Research Ethics in Africa: A Resource for Research Ethics Committees*. Stellenbosch: Sun Press.

Levasseur, Karine. (2000) "Universities and the Regulation of Research Ethics," in Doern, G. Bruce and Stoney, Christopher (Eds.), *Research and Innovation Policy, Changing Federal Government-University Regulations*. Toronto, ON: University of Toronto Press., pp. 242–264.

Mamdani, Mahmood. (2011) "The Importance of Research in a University" (MISR Working Paper No 3).

Ndlovu-Gatsheni, Sabelo. (2020) *Decolonization, Development and Knowledge in Africa: Turning over a New Leaf*. Abingdon, Oxon; New York, NY: Routledge, Taylor & Francis Group.

Medie, Peace A., and Kang, Alice J. (2018) "Power, Knowledge and Politics of Gender in the Global South," *European Journal of Politics and Gender*, 1(1–12), pp. 37–53.

Mite, Kham, and Ndebele, Paul. (2014) "Payment of Research Participants," in Kruger, Mariana, Ndebele, Paul and Horn, Lyn (Eds.), *Research Ethics in Africa: A Resource for Research Ethics Committees*. Stellenbosch: Sun Press, pp. 135–141.

Mudimbe, Valentine. (1988) *The Invention of Africa: Gnosis, Philosophy and the Order of Knowledge*. Bloomington: Indiana University Press.

Musila, Grace A. (2017) "Navigating Epistemic Disarticulations," *African Affairs*, 116(465), pp. 692–704.

Mutenherwa, Farirai, and Wassenaar, Douglas Richard. (2014) "16: Ethics Review of Social and Behavioural Research in an African Context," in Kruger, Mariana, Ndebele, Paul and Horn, Lyn (Eds.), *Research Ethics in Africa: A Resource for Research Ethics Committees*. Stellenbosch: Sun Press, pp. 117–124.

Ndebele, Mwaluko, Kruger, Ouwe Missi Oukem-Boyer and Zimba (2014) "History of Research Ethics in Africa," in Kruger, Mariana, Ndebele, Paul and Horn, Lyn (Eds.), *Research Ethics in Africa: A Resource for Research Ethics Committees*. Stellenbosch: Sun Press, pp. 3–10.

Nyika, A., Kilama, W., Chilengi, R., Tangwa, G., Tindana, P., Ndebele, P., and Ikungura, J. (2009) "Composition of Ethics Review Committees across Africa: Are the Gate-Keepers Rising to the Emerging Challenges," *Journal of Medical Ethics*, 35(3), pp. 189–193.

Nordstrom, Carolyn, and Robben, Antonius. (1995) *Fieldwork under Fire: Contemporary Studies of Violence and Survival*. Berkeley and Los Angeles, CA: University of California Press.

Ogot, Bethwell A. (2009) "Rereading the History and Historiography of Epistemic Domination and Resistance in Africa," *African Studies Review*, 52(1), pp. 1–22.

Obeng-Odoom, Franklin. (2019) "The Intellectual Marginalization of Africa," *African Identities*, 17(3/4), pp. 211–224.

Porter, Elisabeth, Robinson, Gillian, Schnabel, Albrecht, Smyth, Marie and Osaghae, Eghosa (Eds.) (2005) *Researching Conflict in Africa: Insights and Experiences*. United Nations, Tokyo, New York and Paris: University Press.

Sleem, Hany, Abdelhai, Rehab Abdelhai Hamed, Al-Naif, Imad, Gabr, Hala Mansour, Kehil, Et-Taher, Sadiq, Bakr Bib, Yousri, Reham, Elsayed, Dyaeldin,

Sulaiman, Suad, and Silverman, Henry. (2010) "Development of an Accessible Self-Assessment Tool for Research Ethic Committees in Development Countries," *Journal of Empirical Research on Human Research Ethics: An International Journal*, 5(3), pp. 85–96.

Sluka, Jefferey A. (1995) "Reflection on Managing Danger in Fieldwork: Dangerous Anthropology in Belfast," in Nordstrom, Carolyn, and Antonius Robben (Eds.), *Fieldwork Under Fire: Contemporary Studies of Violence and Survival*. Berkeley and Los Angeles, CA: University of California Press, pp. 276–294.

Thomson, Susan, Ansoms, An, and Murison, Jude. (2012) *Emotional and Ethical Challenges for Field Research in Africa*. Basingstoke: Palgrave Macmillan.

Tilley, Helen. (2011) *Africa as a Living Laboratory: Empire, Development and the Problem of Scientific Knowledge*, 1870–1959. Chicago, IL: University of Chicago Press.

Toy-Cronin. (2018) "Ethical Issues in Insider-Outsider Research," in Iphofen, Ron and Tillich, Martin. (Eds.), *The Sage Handbook of Qualitative Research Ethics*. London: Sage, pp. 455–468.

van den Hoonard, Will C. (2018). "The Vulnerability of Vulnerability: Why Social Science Researchers Should Abandon the Doctrine of Vulnerability," in Iphofen, Ron and Tillich, Martin (Eds.), *The Sage Handbook of Qualitative Research Ethics*. London: Sage, pp. 305–321.

Wasuna, Christine, Tegli, Jemee, and Ndebele, Paul. (2014) "Informed Consent in an African Context," in Mariana Kruger, Paul Ndebele, and Lyn Horn (Eds.), *Research Ethics in Africa: A Resource for Research Ethics Committees*. Stellenbosch: Sun Press, pp. 57–62.

Yacob-Haliso, Olajumoke. (2018) "Intersectionalities and Access in Fieldwork in Post-Conflict Liberia," *African Affairs*, 118(470), pp. 168–181.

Wolf, Diane L. (1996) *Feminist Dilemmas in Fieldwork*. Boulder, CO: Westview Press.

Wood, Elisabeth Jean. (2006) "The Ethical Challenges of Field Research in Conflict Zones," *Qualitative Sociology*, 29, pp. 373–386.

5 The application of qualitative methodology to peacebuilding research in Africa

Kenneth Omeje

Introduction

This chapter explores the applications of qualitative research methodology to strengthen policy research and peacebuilding in Africa. One of the key ways of bridging the gap between peacebuilding policies (including neoliberal peacebuilding) and local contexts in Africa is by emphasising the need for peacebuilding policies to be underpinned by context-specific, evident-based research. Over the past decade, there have been diverse donor-funded peacebuilding research outputs and publications in Africa, and most of these studies have shown a predilection for qualitative research methodologies. Many of the emerging research outputs are of questionable methodological rigour and analytical quality, which literally make them to be distrusted in policy circles.

Conventionally, quantitative research seems to be more trusted in international and national development circles mainly because of its empiricist rigour, "scienticization" of claims, quantitative precision and content brevity. This does not imply that quantitative researches are completely error-proof or insulated from methodological biases and abuse. No, they are not, neither is quantitative research superior to qualitative research. Clearly, none can justifiably be said to be superior to the other. Quantitative and qualitative methodologies are just two different orientations to social research, in some limited cases, applicable to different subject matters and contexts, each having its peculiar strengths and weaknesses. However, to enhance their public policy relevance and credibility, there is the need for research works adopting qualitative methodologies to strengthen their game in terms of methodological rigour and presentation.

Qualitative methods of research

Qualitative methods of research are designed to capture social life as people experience it, fundamentally reporting it in a non-numerical form. Qualitative researchers use different types of non-numeric data, e.g. interviews, Focus Group Discussions (FGDs), surveys, field observations and assorted secondary data (journal articles, webpages) in order to explore and evaluate social phenomena (Garcia, n.d.). The methods of analysis are mixed, flexible and creative, including methods such as descriptive narratives, content analysis, thematic analysis, discourse analysis, grounded theory (GT) and theory application/analysis. In their philosophical orientation, qualitative methods are generally associated with post-positivism. Quantitative methods, on the other hand, are used in collecting and analysing data that are capable of being expressed in numerical values. Triangulation across the two methods is possible, practicable and indeed highly recommended. It is an advantage for a researcher to have adequate knowledge of both qualitative and quantitative methodologies, and also be skilled in their contextual applications.

Major qualitative methods of fieldwork data collection commonly used in peacebuilding research

There are diverse methods of qualitative research, but in this chapter it will suffice to concentrate on the key methods that tend to be frequently used in contemporary peacebuilding research, namely interviews, observation, FGD and ethnographic method. The listed methods in this chapter is by no means exhaustive. Qualitative methods can be used to generate various types of data, the most commonly prevalent being the following types: in-depth/unstructured interviews, including video, semi-structured interviews, structured interview schedule containing substantial open comments, focus groups, unstructured or semi-structured diaries, observation field notes/technical fieldwork notes, case study notes, minutes of meetings, press clippings and court transcripts (see ICPSR, n.d.).

Interviews

The qualitative research interview is a conversation between parties to an investigation or research (the researcher and the researched) on issues supposedly of mutual interest (Kvale, 2008). It is a professional kind of conversation through which the researcher gets to know about other people: learn about their experiences, feelings and about their perceptions of the world around them. The purpose of the research interview data is to help the researcher to construct meaningful stories (researcher's narrative) about a research problem based on his/her understanding and interpretation of respondents' narratives. In terms of format, interviews can take the form of one-to-one or group interview.

There is unequal discursive power relation between interview parties – the interviewer and the interviewee or the researcher and the researched. The interviewer invariably has the upper hand in the interview power relations because she/he is the one that tells the final story based on his or her subjective understanding and interpretation of the respondent's narratives. Potentially, the interviewer has the liberty and power to screen and sift from the respondents' narratives the precise interview data or responses to use in his or her data interpretation/analysis, in what form the data could be used, to what extent or proportion and most importantly, the meanings to assign to the employed set of data. Research is a human activity or occupation, and as such disingenuous researchers could easily use their marked discursive power advantage to overly select, twist, and amplify data with the goal of producing a desired result. It is because of this seemingly undue discursive power advantage in the hands of the interviewer/researcher that many mainstream empiricists inclined to quantitative research tend to view studies based exclusively on qualitative methodology with strong reservation and suspicion.

In order to mitigate the potential for an abusive use of qualitative research interview (and indeed, qualitative data in general), many critics recommend safeguards such as triangulation, rigorous training in data stewardship and data coding and processing using computer-assisted qualitative data analysis software (e.g. NVivo, MAXQDA, ATLAS.ti) as well as archiving of qualitative data for the purposes of secondary analysis, verification, comparison with other data and interrogation by other researchers. Incidentally, most of the qualitative data archiving resource centres and services are available in the developed countries. Many international organisations and UN specialised agencies provide restricted services to archives of social science data in general, which include datasets on specific African countries.

There are three main types of interviews, namely: *Structured Interview*: In structured interviews, the interviewer puts relatively strict structure to the format of questions asked and answers elicited. *Unstructured Interview*: This type of interview has no rigid form. The interviewer allows conversation to flow naturally without any restrictions to how she/he probes on issues and how the interviewee responds to the questions. *Semi-Structured Interview*: This type of interview combines some of the features of the earlier two methods (structured and unstructured), but tries to keep a healthy balance.

Some of the important considerations in planning a research interview are: clarity about the type of interview to be held (structured, unstructured, or semi-structured); need to obtain the consent of the interviewee; arrangement to take field interview notes, i.e. during or immediately after the interview depending on the nature of research; and decision on whether or not one needs electronic facilities like tape recorder, camera or video equipment. It is important to note that it is both unethical and unlawful (in most countries) to use any electronic device to record someone's interview without the prior informed consent and permission of the respondent. Other essential considerations for using the interview method in fieldwork data collection are: suitability of the interview location, venue and timing; determining the language to be used in the interview based on a clear knowledge of one's research audience's socio-demographic background; and decision about whether research assistants are required to help in conducting the interviews, modalities for their recruitment, training and remuneration as well as the operational techniques and etiquette for the actual interview process.

Observation method

Observation is about a researcher's eyewitness account of what people do as opposed to what they say or report (see Brewer, 2013). Depending on the nature and duration of one's research, observation can be episodic, repeated, periodic or prolonged. Episodic observation is usually one-off and tied to a specific event or phenomena. In repeated observation, an event is observed many times to ascertain the degree of regularity, consistency, comparability or replicability. Periodic observation usually occurs in regular or predictable intervals. The rationale is to establish the degree of regularity or comparability. Prolonged observation is long-term based, and it is a method that employs repeated or periodic observing of events and phenomena. In repeated, periodic and prolonged observation, the researcher should create an observation schedule to document and compare results. In addition, a field note is generally required in all forms of research fieldwork.

Fieldwork observation occurs in the natural setting of research. The method depends heavily on the researcher's perception and interpretation of what she/he sees which may have certain advantages and disadvantages. To enhance accuracy and reliability in perception and interpretation of observation, it is important that the researcher is sufficiently conversant with the cultural values, ethos, symbols and practices of the people. Follow up inquiries are important for probing and seeking clarity on any subterranean or ambiguous fieldwork observations.

The origin of observation method in the liberal sciences is closely associated with the works of some of the earliest colonial and Victorian anthropologists, sociologists and historians in their study of comparative social structures – scholars such as Sir Henry Maine (1822–1888), John Ferguson McLennan (1827–1881), Sir Edward Tylor (1932–1917) and Sir James Frazer (1854–1941). The approach of these pioneer scholars was vitiated by a paradigm of "ethnographic distance" still significantly prevalent in contemporary ethnographic studies, a social distance linked to the fact that these Western ethnographers were "outsiders" observing "others" in their natural settings. This method of "outsiders" collecting observational data to produce knowledge from the subjective experiences of other people raises some epistemological questions about the veracity of such knowledge.

Observation method can be used in various types of study but specifically the method is considered best suited for (i) studying sensitive subjects in which people are not too willing or comfortable to talk or answer questions; and (ii) studies that have abundant time for fieldwork, in which case repeated and prolonged observation would be recommended.

Participant and non-participant observation

There are two broad types of fieldwork observation: participant and non-participant observation. In participant observation, the researcher participates directly in the daily life of the people or group she/he is studying (Lofland, 2006). The researched group may or may not know that the researcher is deliberately studying them. When the researched community are aware that the observer is studying them as a participant, then we have what is known as obtrusive participant observation. When the opposite is the case, i.e. the researched are unaware that they are being studied and observed, we call this unobtrusive participant observation. There can be *total* or *partial* participation. In total participation, the researcher takes full unrestrained part in the phenomenon being studied, which would mean that in a sensitive research such as "the study of commercial sex workers' responsiveness to government crackdown of red-light districts and nude night clubs," the researcher volunteers to take active part in all the activities of the sex workers. There are obvious personal risks and moral issues involved in this sort of unrestrained participatory study. The opposite and more measured type of participatory research is partial participation in which the researcher *ab initio* determines his or her degree of participation, being cognizant of the risks and hazards of unstinted participation.

It is pertinent to remark that obtrusive participant observation is potentially problematic as the audience can be reactive to an investigator's observation, a phenomenon known as the Hawthorne Effect in social research. The Hawthorne effect is a term generally used to describe a research situation in which observed subjects deliberately modify their behaviours because they know they are being studied and thus want to put up their best performance (Cherry, 2017). It is a condition more likely to occur when an "outsider" is obtrusively using observational method to study a small close-knit social group involved in subcultural practices viewed by the larger society as obnoxious or unlawful. In such a circumstance, some of the stigmatised or prohibited aspects of the cultural practices could be sanitized, concealed or minimized by the observed group in order to make them appear less reproachable. A typical example of studies that could potentially be susceptible to the Hawthorne effect would be ethnographic studies of undergraduate students' secret cults known to be widespread in Nigerian universities and mostly associated with gun violence, rape, talismanic power, use of hard drugs and violent crimes as well as clandestine fetish ceremonies and voodoo rituals carried out in obscure forests at midnight hours.

In non-participant observation, the researcher observes the studied group as an outsider, without participating in their activities. We also have obtrusive and unobtrusive non-participant observation, that is, corresponding to whether or not the researched community knows that the non-participant observer is studying and observing them.

Focus Group Discussion

FGD can be defined as a group of individuals carefully composed by a researcher to discuss a specific research topic or certain issues that are supposedly of mutual interest to the researcher and the participants or discussants (see Morgan, 2013). FGD topics can be wide-ranging, but they mostly comprise issues of opinion, attitude and experience. Using the FGD method requires specialist skills and the researcher has to ensure that the discussion is led by a trained moderator, supported by a time keeper/note taker. FGD is a moderately structured kind of discussion, albeit less structured in exploratory research – or what could be termed "naturally occurring group" discussion. As a relatively structured kind of discussion, the key issues around which FGD is structured are: topics/questions of discussion, size of participants, venue of discussion, sitting arrangement and time allowed for contribution.

Best recommended practices in contemporary social research suggest that to be effective, a standard FGD should comprise between 8 and 12 participants and lasts for about one to two hours. On some sensitive subjects and among certain cultures, it is advisable to reflect certain sensitivities in planning FGDs, e.g. gender (having separate FGDs for males and females), age (separate FGDs for specific age cohorts), social status (separate FGDs for audience of different social status on a sensitive social issue) and so forth. The FGD method is especially suitable for minors and illiterate or semi-literate populations.

FGDs are often preceded by sensitization to educate potential participants on the purpose of the discussion and reassure them about ethical issues. Sensitization is best done during participants' recruitment. FGD is usually started off with self-introduction by participants, followed by the moderator highlighting the purpose, format and procedure of the meeting.

Among the important considerations in planning FGDs are:

a Clarity about the purpose and the effectiveness of the method in generating the required type of data. In other words, is the method the right method for the subject of inquiry?
b Determination of how many FGDs you require for the study and where/when to hold the meetings.
c Identify the relevant category of participants required for the FGD and how to select discussants or participants.
d Determine who should preside over the FGD, usually two persons – a moderator and a recorder (i.e. note taker/time keeper).
e Determine whether or not you need to use electronic facilities for the FGDs – audio or video recorder, camera and so on. You would need the prior consent of participants to use any electronic storage devices.
f Decision about the specific language for holding the FGD and also the language for reporting it. Depending on the composition of the audience, an FGD could be held in a local African language used by all the participants and reported in English or any other national lingua franca.
g Decision on whether to pay the FGD participants a stipend (for their travel expenses/inconveniences) and how much. It is advisable to offer

some modest stipend to participants (preferably a pre-agreed amount), especially in a circumstance where they are likely to incur some financial expenses or logistical inconveniences to get to the FGD venue.

h Provision of refreshment (e.g. snacks, tea/coffee).

FGDs should be moderated with courtesy and good facilitation skills. The moderator has to ensure that the session is not dominated by the most gifted communicators. It is necessary to acknowledge everyone's contributions with verbal appreciation or positive body language. FGD report is more or less written like minutes of a business meeting, but can be more detailed than the latter. A separate report is produced for each FGD. The report is prepared by the note taker and is developed from the discussion notes. The report should have the list and basic biodata of all participants. This can be obtained before or after the discussion session or perhaps at the point of recruitment of participants. The report narrative is written in reported speech against the names of contributors. Judgment is required in compiling the final report. Comments considered irrelevant could be weeded off.

Some of the worst practices in the use of FGD in social policy research include having too many questions; inviting or involving too many participants; absence of ground rules and non-enforcement of rules in the FGD process; poor time management resulting in overly prolonged sessions; poor recording and reporting of data; and socio-cultural insensitivity, such as in the composition of participants and fielding of questions.

In most peacebuilding researches, FGD has often been used as a complementary method of primary data collection (often complementing the interview method). In such a context, it will be inadvisable to pose in an FGD session questions that could be better captured by in-depth interviews. Sensitive personal questions relating to people's nightmarish and traumatic experiences are, for instance, better captured using in-depth interviews.

Ethnographic method

Ethnographic method is used in studying socio-cultural phenomena, and in using the method the researcher tries to study and understand the phenomena as much as possible from the point of view of the subjects being studied. It involves researchers "immersing" themselves in the culture or social field of the studied group for a long period of time (British Library, n.d.). The ideal duration of an ethnographic survey is heavily contested, but depending on whether the ethnographer has prior acquaintance with the socio-cultural setting or not, it could range from six months to a couple of years and perhaps much longer. The rationale is to sufficiently observe, learn and document the group's cultural patterns, social relationships and the meaning they attach to the things they have or do (ibid.; see also O'Reilly, 2005).

What is popularly known as the ethnographic method is an amalgam of different qualitative methods (in-depth interviews, oral history, participant observation). The modern ethnographer is trained to study a group from the group's perspective, which is contrary to the largely ethnocentric approach adopted by most Victorian anthropologists that originated the ethnographic method through their works in various small-scale societies in the global South. The ethnographic method is widely used in anthropology and sociology to study micro-groups, criminalised sub-cultures and small-scale societies (ibid.; see also Barker, 2001).

Qualitative data analysis methods applicable to social policy and peacebuilding research

Let me in the following section briefly discuss some of the key methods employed in discussing, interpreting and analysing qualitative data applicable to peacebuilding and social policy research in general.

Excerpts

An excerpt is a piece of discourse or quote extracted from a larger body of information. Excerpts can be taken from a broad range of data in research. Excerpts, like any other qualitative data, are not self-explanatory in research. Meanings are assigned through the prism of the researcher's interpretive discourse.

Let's take the following example of an excerpt developed from an in-depth interview question conducted during a 2013 APN research on "the Role of NGOs in Peacebuilding in South Sudan." The respondent answered *inter alia*:

ACROSS has set up programs in Yei and Rumbek aimed **at training local communities to use natural medicine**. We have trained them on the effects of **polygamy and its impacts on HIV/AIDS**. We have given **civic education on the negative effects of cattle rustling by organizing sports days to promote peacebuilding** across the ten States of South Sudan. **[...]**We also have a long-term initiative aimed at resettling returnees in South Sudan. **We work with the United Nations High Commission for Refugees (UNHCR) to put them in camps**. We assist them with basic necessities and we have done this for about 2,000 returnees so far. The assistance not only involves shelter but services, such as building schools and digging of water boreholes, which has enabled them to be reintegrated back into their communities. **[...]**We have also set up **long-term peace education programs** in Lakes and Central Equatoria States which aimed at constructive peacebuilding agendas. We have further been involved in **long-term teacher training**. Last December [*December, 2012*] 260 teachers that we financed graduated with Primary School Teaching certificates in the different States of South Sudan, including some areas of North and South Kordofan.

The challenge that many researchers face is how to interpret excerpts like the preceding example, and as such they sometimes simply reproduce them as an indented quote and proceed with their narrative. Excerpts such as the foregoing could, for instance, become the basis for further qualitative interpretation using methods such as thematic analysis, content analysis and discourse analysis. The phraseologies and incomplete sentences that I have highlighted in bold are constructs from which themes could, for example, be developed to guide the researcher in further discussion and analysis.

Shorter excerpts of one to three lines could for instance be cited as quotes to support, strengthen, interrogate or challenge an argument, especially if the expressed view is a widespread discourse that cuts across many respondents' narratives. Longer excerpts could be potentially used by the researcher to develop thematic constructs for further application in engaging and interpreting data. It is noteworthy that the three dotted lines in the earlier excerpt [**...**] indicate a deliberate exclusion or omission of some of the respondent's original discourse considered inappropriate, irrelevant or overly sensitive to be reproduced or placed in public domain.

Paraphrasing

Paraphrasing is when a researcher uses his/her own words to articulate and express the idea of someone else such as a piece of information, sentence or passage. The main rationale for paraphrasing is to enhance clarity and logical fitness of the original information to the researcher's narrative. There are two major conditions for paraphrasing in research. The first is that the researcher must adequately understand the original text or information, which is important to avoid any misrepresentation. The second is that the paraphrased information or idea must be credited to its original source, which necessitates some form of acknowledgement and referencing. There is a fine line between paraphrasing and plagiarism. The following example of an original text and a paraphrased version will help to further illustrate the fine line between a paraphrased idea and plagiarism.

Original text

"The political independence Europe 'granted' to her overseas territories in sub-Saharan Africa during the 1960's changed nothing... In the contrary, Europe's economic domination over sub-Saharan Africa deepened" (Rodney, 1972, p. 149).

Paraphrased version

Sub-Saharan Africa's political independence from Europe during the 1960s changed nothing in Africa; but, on the contrary, it deepened Europe's domination of the region (Rodney, 1972, p. 149). In practice, many researchers will take the original text and get it modified in what I have represented as the paraphrased version as their own idea, without any in-text citation or acknowledgement of the original source. This is a case of subtle plagiarism. Paraphrasing should be kept short and if longer than four lines should be indented as a lifted direct quote.

Typology

In social research, a typology is the classification of phenomena or things according to their major types and observable characteristics. It is the outcome of a grouping process in which categories are constructed based on the nature and levels of similarities and differences amongst constituent elements. A researcher can construct typology out of qualitative data or apply existing relevant typologies to the analysis of data.

Below is an example of a context-specific typology constructed from data classification using a 2013 APN-funded research. Data for this typology were collected using interview methods (Table 5.1).

Table 5.1 Respondents' Views on Priority Areas for Post-Conflict Peacebuilding

Thematic Issues	*Empirical Indicators and Manifestations*
1a Border disputes (Internal) **1b Border disputes (External)**	• Jonglei, Warrap & Lakes States disputes • Warrap / Unity States disputes • Border disputes between South Sudan and Sudan (status of the oil-rich Abyei, Blue Nile and South Kordofan)
II Peace-building and reconciliation	**Perceived need for:** • Unity among South Sudanese • Rebuilding social fabric through healing and reconciliation • Inclusivity and equitable treatment of the citizens from all ethnic backgrounds • Reconciliation among antagonistic communities and ethnic groups
III Security sector reform and democratization	**Perceived need for:** • De-militarization • Disarmament of armed civilian population • Rule of law and transparency • Improvement in public safety and security • Tackling the issue of cattle rustling in various states (inter-ethnic violence) • Dealing with armed militia groups within the country (ethnic insurgencies)
IV Development infrastructures and social amenities	**Perceived need for:** • Provision of development infrastructures like clean water, tarmacked roads, electricity and educational institutions • Housing infrastructure (most critical for IDPs and some post-war returnees) • Service delivery such as public health and medical care, facilities in public schools, etc.
V Employment creation, especially for the youth	**Perceived need for:** • Creation of job opportunities for citizens (youth) • Vocational skills training (youth and women)

Source: Omeje & Minde (2013)

From the earlier typology, each of the thematic issues ("I" to "v") could form the basis for a detailed narrativized discussion of the respondents' views on priority areas for post-conflict peacebuilding, and the column on the empirical indicators/manifestations will provide the analytical rubrics to guide the discussion.

In addition to developing your own typology from your fieldwork data, a researcher could import a relevant existing typology from the literature as a framework to interpret or interrogate primary and secondary data. An example of an existing typology is Johan Galtung's threefold typology of violence which includes direct violence, structural violence and cultural violence (Ramsbotham, Woodhouse, & Miall, 2005). You can only apply a typology you consider to be relevant to your research and such a typology has to be clearly outlined and critically discussed.

Content analysis

Content analysis is concerned with determining the occurrence of certain concepts or key words in a text and analysing or interpreting them either in isolation (i.e. conceptual analysis) or in relation to other associated concepts (relational analysis). Meaning is figured out contextually. In content analysis, "text" is broad-ranging and includes: the content of information, messages, written texts of any kind (e.g. books, newspapers and memos), interview, speeches, conversation of any kind and so forth.

Conceptual analysis is applied to establish the frequency of occurrence of certain concepts (words or phrases) in a text and to analyse their various meanings. The concepts can include synonyms, e.g. in a research on "political violence" the researcher might be interested in the following words and phrases in a text: war, warfare, insurgency, fighting, violent attack, violent behaviour, hostility, aggression, insurrection, low-intensity warfare. The researcher is particularly interested in how many times a concept occurs and the consistency in meanings as well as in use of synonyms and their meanings.

Relational analysis is applied to show both the meaning of concepts and how they are related to one another in a text. The researcher is interested in how a particular concept is used in relation to another, the frequency of their simultaneous occurrence and the meanings they convey. For example, in studying "political violence" in relation to "deprivation," a concept such as "deprivation" could be stretched to include synonyms like "denial, withdrawal of services or benefits, lack, dispossession, scarcity, etc." In an in-depth interview-based research, counting of the frequency of concepts could be done across all the interview transcripts to aid both conceptual and relational analyses. In this case, the use of computer-assisted qualitative data analysis software is considered vital. Content analysis is similar to thematic analysis in many ways save that content analysis is more amenable to the application of quantitative (statistical) methods of analysis.

Thematic analysis

This is an analytic model based on reviewing, coding and categorising of qualitative data with a view to discovering patterns of meaning and developing analytical themes. The coding uses concepts and phrases (themes) to depict or label sections of data. Coding concepts and theme development could be derived inductively from the researcher's reading of the content of data or deductively by applying an existing relevant concept to a set of data. Coding could also be driven by existing theoretical and essentialist constructs evident in the data or on some latent assumptions underpinning data. Ultimately the purpose of thematic analysis is to identify patterns of meaning across a dataset that provide an answer to the investigated research question (Braun & Clarke, 2006). Once themes are developed and data categorised, the way they are used and analysed depends on the intentions of the researcher. Thematic analysis can be a preliminary step to some more fine-grained analysis using other qualitative methods such as discourse analysis, GT and interpretivism.

Theory application

Researchers can apply any of the existing social and liberal sciences theories in isolation or in combination either as a framework of analysis or for data interpretation and interrogation. There is inexhaustible number of theories in the social and liberal sciences that could be potentially applied in the study of any social phenomenon. Theories need to be evaluated and applied based on their relevance to a specific research topic or question as well as the researcher's informed knowledge of their explanatory powers and limitations.

Some of the important requirements for theory application include knowledge of the sociological origin and key assumptions of the theory, direct familiarity with the works of the chief proponents of the theory and knowledge of the merits and explanatory strength of the theory. Other important requirements are familiarity with any outstanding studies that have applied the theory and knowledge of the criticisms and limitations of the theory. Another factor that might be important is the knowledge of how to use triangulation to remedy the shortcomings of any particular theory or framework of analysis.

Discourse analysis

Discourse analysis is a multifaceted theory and methodology of research that has "ambiguous" conceptions and connotations even from the standpoint of many of its ardent proponents. In general terms, discourse analysis can be defined as the interrogative analysis of the content of language, written texts, theories and institutional structure/behaviours, with a view to revealing their underlying power relations, assumptions and the interests they tend to maximize in society. In conceptual and methodological terms, discourse analysis subsumes the various strands of post-positivist or post-structuralist approach to investigation and analysis.

From a post-structuralist view of social reality, the famous French philosopher and one of the most noted pioneers of discourse analysis, Michel Foucault, defines discourse as:

> ways of constituting knowledge, together with the social practices, forms of subjectivity and power relations which inhere in such knowledges and relations between them. Discourses are more than ways of thinking and producing meaning. They constitute the 'nature' of the body, unconscious and conscious mind and emotional life of the subjects they seek to govern.
>
> (Weedon, 1987, p. 108)

The Foucauldian concept of discourse sets up a constitutive relationship between meaning and power in social practice; every move to meaning-making comes about from a position of power -i.e. power both structuring and structured by the social positions available within the practice (Chouliaraki, 2008, p. 1). Discourses are shaped by contextual power relations, and they invariably set the context and limits of social knowledge.

In discursive methodology, social representation does not have a fixed meaning but "surplus meaning," and reflexivity is at the centre of meaning construction in research. Reflexivity is concerned with the notion that the researcher's subjectivities are central to the type of knowledge constructed or produced in research. It is "an awareness of the researcher's contribution to the construction of 'meanings' throughout the research process; an acknowledgment of the impossibility of remaining 'outside of' one's subject matter while conducting research" (Willig, 2001, p. 10). In reflexivity, a researcher "explores the ways in which his involvement with a particular study influences, acts upon and informs the research process and outcome" (ibid).

The discursive framework is interested in disguised and hidden representations or meaning (e.g. the issue of symbolisms, rituals and taboos). These important aspects of language and culture help in understanding a people.

In discourse analysis, social phenomena are culturally specific or constituted, and culture is broadly conceived not as an objective reality but as resources of meaning-making, language and image, which are available for use in a community of social actors at any given time (Chouliaraki, 2008, p. 1). Social meaning or meaningful reality is socially constructed and language is central to social interaction and the construction of meaning (even in research).

Some relevant keys in discourse analysis

i Adopt a critical approach to social knowledge, interrogating all received knowledge to understand how and why they were constructed and for what purpose. Our knowledge of the world is historically and culturally specific and cannot be treated as an objective truth.

ii Understand and account for the link between social actors, social processes and social action; identify and analyse social actors, the objects on which they are perceived to be acting, why they undertake certain actions as well as the underlying power relations (real or assumed).

iii Explore power configurations; specifically, how power is spread across different social practices. In this regard, identify and analyse what actors do using the resources or power at their disposals, why they do it and the limitations they face.

iv Interrogate any real or perceived challenges in power relations, the changes that occur in power dynamics as well as the implications of any observed changes for actors' empowerment and disempowerment.

v Identify discourses and unravel meanings through reflexive and deconstructive interrogation of data (e.g. interview narratives and written texts). You must note that every claim or view expressed by someone regarding a social phenomenon is a discourse (Cutting, 2002; Sage, 2002).

Discourse analysis is a method used in different fields of humanities and the social sciences.

Triangulation

Triangulation in the social sciences is an attempt to explore social phenomena and also validate research data from multiple standpoints, usually three or more standpoints (Kennedy, 2009). There are four methods of triangulation: (1) data triangulation (involving data that traverses time, place and persons), (2) methodological triangulation (multiple data collection methods), (3) investigator triangulation (involving multiple investigators) and (4) theory triangulation (application of multiple theories). Competently and professionally applied, triangulation has the advantage of enriching research and minimizing bias.

Grounded theory

GT is a research method that seeks to develop theory that is grounded in systematically gathered and analysed qualitative data, usually (ethnographic) interview, observation or FGD data. GT is mostly used in qualitative research requiring data coding using appropriate conceptual constructs or themes. Conceptual constructs for coding are often selectively (refined) derived from data (axial coding); concepts can also be coined or adapted from literature. GT was originated in the 1960s by two American sociologists – Barney Glaser and Anselm Strauss. Methodologically, it adopts an empirico-positivist approach different from mainstream positivism. GT aims to understand research and develop theory from the perspective of the researched audience or sample. It is appropriate for studies that aim to explain a process, not to test or verify an existing theory (Schatzman & Anselm, 1973). GT is for the most part problem-solving oriented.

Brainstorming exercise on grounded theory

a Develop a GT (*thematic building blocks only*) of internal conflicts and peacebuilding in the north-eastern region of South Sudan from an in-depth interview excerpt in response to the following question:
b What do you consider to be the most important priority area for building sustainable peace in the north-eastern region of South Sudan?

The respondent's narrative is as follows:

Well, I feel that the most important priority area for government and other stakeholders should be **to find a way of addressing internal conflicts between different communities and States**. Warrap State, Unity State and Lakes State experience **too much violence due to cattle raiding** and there has been **an increase in inter-state cattle rustling in this region**. I recently visited a particular Nuer village in Unity State where the local people have suffered a lot in the hands of their Dinka neighbours in Warrap State. Their village was attacked and burnt down by their Dinka neighbours with all their animals taken away. **Many were maimed and killed and a large number of survivors, particularly women, children and elderly men, ended up in IDP camps**. The young men fled to a neighbouring Nuer village from where they planned and executed a **revenge attack on their adversaries. It becomes an endless cycle. Cattle raiding persist in the region because of poverty and the too many young people having firearms**; and also **because of revenge attacks between communities, lack of jobs, and competition over grazing land and water**. Young men also indulge in cattle raiding to support one another because **marriage bride price is paid using cattle**. Despite the **efforts by the government to disarm people** in this region, **small arms and light weapons are still in large numbers**, and this further contributes to cattle rustling and continuous warfare between communities. The government is, however trying its best to deal with cattle rustling. **They have made it a crime** and **measures aimed at establishing joint inter-State courts along the borders of the three States have been agreed upon**. The **government has also resolved to deploy joint mobile forces in hotspot areas** along the common borders while **State governors will conduct regular exchange visits in the areas. Reports regarding the border conflicts in Warrap, Unity and Lakes States are well known; they have been published by the international agencies working in these areas of border conflicts**, for example UNMISS, MSF. **It is difficult for us to work and distribute**

food and other humanitarian services in Jonglei due to cross border clashes brought about by cattle rustling.[1]

The earlier narrative is typical of how many interviewees responded to the corresponding question, and I have highlighted in bold some of the themes, phraseologies and incomplete sentences (a set of flexible thematic constructs) from which a researcher could possibly develop the relevant conceptual constructs for a GT. The highlighted flexible thematic constructs need to be further synthesised using suitable conceptual constructs developed from the data, coined by the researcher or imported from relevant literature. Subsequently, the delineated conceptual constructs have to be arranged in a chronological order to form building blocks for the GT. Furthermore, the conceptual constructs can become the basis of systematic coding of data using the various narratives provided by the respondents.

Below are some of the flexible thematic constructs from the raw data and the corresponding conceptual constructs that could be used for data coding/GT.

From Table 5.2, the refined conceptual constructs can become the basis of systematic coding of data and GT using the various narratives provided by the various respondents. New conceptual constructs could be added as the need arises.

Table 5.2 Brainstorming Exercise on Grounded Theory: A Sample Framework for Thematic Coding

		Flexible Thematic Constructs from Raw Data	*Refined Conceptual Constructs for Data Coding/GT*
1	I	… to find a way of addressing internal conflicts between different communities and States.	Nature of the conflict
	II	… too much violence due to cattle raiding	
	III	… an increase in inter-state cattle rustling in this region.	
2	i	Many were maimed and killed and a large number of survivors, particularly women, children and elderly men, ended up in IDP camps.	Effects of the conflict
	ii	It is difficult for us to work and distribute food and other humanitarian services in Jonglei due to cross border clashes brought about by cattle rustling.	
3		… revenge attack on adversaries and the conflict becomes an endless cycle	Dynamics of the conflict

(*Continued*)

		Flexible Thematic Constructs from Raw Data	*Refined Conceptual Constructs for Data Coding/GT*
4	a	Cattle raiding persist in the region because of poverty and the too many young people having firearms;	Factors for persistence of the conflict
	b	… because of revenge attacks between communities, lack of jobs, and competition over grazing land and water.	
	c	… marriage bride price is paid using cattle.	
	d	small arms and light weapons are still in large numbers	
5	a	… efforts by the government to disarm people.	Government's conflict intervention strategies
	b	They have made it (cattle rustling) a crime	
	c	… measures aimed at establishing joint inter-state courts along the borders of the three States have been agreed upon.	
	d	… government has also resolved to deploy joint mobile forces in hotspot areas.	
	e	State governors will conduct regular exchange visits in the areas.	
6	a	Reports regarding the border conflicts … are well known; they have been published by the international agencies working in these areas of border conflicts,	Involvement of NGOs/international agencies
	b	It is difficult for us to work and distribute food and other humanitarian services in Jonglei due to cross border clashes brought about by cattle rustling	

Source: Developed by the author.

The main philosophical debate about GT is on whether research should be theory-testing or theory-generating, because GT is essentially aimed at providing empirical explanation to a puzzle with the object or possibility of developing a flexible "Middle-Range Theory." Middle-range theories, first described by the American Sociologist Robert K. Merton, are theories "that transcend sheer description or empirical generalisation" by proffering "limited sets of assumptions from which specific hypotheses are logically derived and confirmed by empirical investigation" ... "in their efforts to develop a unified theory that will explain all the observed uniformities of social behaviour, social organization and social change" (Merton, 1967, pp. 39, 68). One of the major criticisms of GT is that because it aims to construct theory from data, GT blurs the distinction between empirical data and the phenomenon requiring explanation.

Conclusion

In this chapter, I have explored the practical dimensions and challenges of peacebuilding in Africa as well as the need for promoting evident-based policy research in peacebuilding on the continent. This chapter has tried to elucidate the imperatives of qualitative research methodology – types of data, methods of data collection and methods of data interpretation/analysis – including how they could be contextually applied to strengthen social policy research and peacebuilding in Africa.

An important aspect of qualitative research methodology that has not been discussed in this chapter for lack of space is the issue of population and sample selection in fieldwork-related research, in particular the techniques of non-probability sampling considered more relevant to the post-positivist paradigm mostly favoured by qualitative researchers. Fieldwork researchers are encouraged to be sufficiently acquainted with non-probability sampling – its various methods and application. Some of the key methods of non-probability sampling are convenience sampling, accidental sampling and purposive sampling; and quota sampling, snowball sampling and sequential sampling (see Saunders, Lewis, & Thornhill, 2012). Judgment about the adoption and application of any sampling method has to be carefully made based on the nature of the research and the investigator's knowledge of the research audience and fieldwork terrain.

Finally, it is pertinent to remark that in qualitative research, data can be extracted and classified and reports created manually or by use of computer-assisted software such as Nvivo, Atlas, MAXQDA and so forth. Interested scholars and researchers could refer to the following YouTube website for a Nvivo fresh-learner friendly tutorial: https://www.youtube.com/watch?v=Xe0NpJPLQ6k (accessed on 12 January 2018).

Note

1 Source: Omeje and Minde (2013). This narrative is a deliberately distorted and amplified version of the interview response of a female respondent working with a foreign NGO in South Sudan. The interview was held in Juba on 29/01/2013 and I have deliberately distorted and amplified the narrative to suit the purpose of this GT brainstorming exercise.

References

Barker, Chris. (2001) *Cultural Studies and Discourse Analysis: A Dialogue on Language and Identity*. London: Sage.

Braun, Virginia & Victoria, Clarke. (2006) "Using Thematic Analysis in Psychology." *Qualitative Research in Psychology*, 3(2), pp. 77–101.

Brewer, John. (2013) "The A-Z of Social Research Observation, Overt and Covert," Sage Research Methods, http://www.uk.sagepub.com/gray3e/study/chapter16/Encyclopaedia%20entries/Observation_overt_and_covert.pdf (accessed 15 October 2017).

British Library. (n.d.) *Social Science Collection Guides: Topical Bibliographies*, http://www.bl.uk/reshelp/findhelpsubject/socsci/topbib/methsocial/methodsresearch.pdf (accessed 10 November 2017).

Cherry, Kendra. (2017) "What Is the Hawthorne Effect?" *Very Well Psychology Glossary*, 27 July, https://www.verywell.com/what-is-the-hawthorne-effect-2795234 (accessed 8 January 2018).

Chouliaraki, Lelie. (2008) "Discourse Analysis." LSE Research Online, http://eprints.lse.ac.uk/21564/1/Discourse_analysis_(LSERO_version).pdf (accessed 8 September 2017).

Cutting, Joan. (2002) *Pragmatics and Discourse*. London: Routledge.

Garcia, Ana. (n.d.) "Module 1 – Nvivo Basics: Getting Started." University Library, La Trobe University, Melbourne, Victoria, https://www.youtube.com/watch?v=Xe0NpJPLQ6k (accessed 13 September 2017).

ICPSR. (n.d.) "Guide to Social Science Data Preparation and Archiving – Phase 3: Data Collection and File Creation," https://www.icpsr.umich.edu/icpsrweb/content/deposit/guide/chapter3qual.html (accessed 16 September 2017).

Kennedy, Patrick. (2009) "How to Combine Multiple Research Options: Practical Triangulation," http://johnnyholland.org/2009/08/20/practical-triangulation (accessed 8 September 2017).

Kvale, Steinar. (2008) *InterViews: An Introduction to Qualitative Research Interviewing* (2nd Edition). London: Sage.

Lofland, John. (2006) *Analyzing Social Settings: A Guide to Qualitative Observation and Analysis* (4th Edition). London: Wadsworth.

Merton, Robert K. (1967) *On Theoretical Sociology: Five Essays, Old and New*. New York: Free Press.

Morgan, David L. (2013) 'Focus Groups as Qualitative Research: Planning and Research Design for Focus Groups,' Sage Research Methods, http://www.uk.sagepub.com/gray3e/study/chapter18/Book%20chapters/Planning_and_designing_focus_groups.pdf (accessed 14 September 2017).

Omeje, Kenneth & Minde, Nicodemus. (2013) "Peace-Building and State-Building in a Fragile Rentier State: The Case of South Sudan." Report of Empirical research project funded by African Peacebuilding Network (APN) of the Social Science Research Council (SSRC), Washington, DC, August.

O'Reilly, Karen. (2005) *Ethnographic Methods*. London: Routledge.

Ramsbotham Woodhouse O.T. & Miall, H. (2005) *Contemporary Conflict Resolution*. Cambridge: Polity Press.

Rodney, Walter. (1972). *How Europe Underdeveloped Africa*. London: Bogle-L'Ouverture Publications.

Sage. (2002) "The Field of Discourse Analysis," Sage Publications Online, https://uk.sagepub.com/sites/default/files/upm-binaries/9452_011011Ch1.pdf (accessed 8 December 2017).

Saunders, Mark, Lewis, Philip & Thornhill, Adrian. (2012) *Research Methods for Business Students* (6th Edition). London: Pearson Education Limited.

Schatzman, Leonard & Strauss, L. Anselm. (1973) *Field Research: Strategies for a Natural Sociology*. Englewood Cliffs, NJ: Prentice-Hall.

Weedon, Chris. (1987) *Feminist Practice and Poststructuralist Theory*. Cambridge, MA: Blackwell Publishers.

Willig, Carla. (2001) *Introducing Qualitative Research in Psychology: Adventures in Theory and Method*. Buckingham: Open University Press.

6 Quantitative research methods in field-based peacebuilding

Olayiwola Erinosho

Introduction

This chapter is about researching in peacebuilding using time-tested quantitative methods. It starts with the clarification of some key concepts, namely *peacebuilding, field based research*, and *quantitative research methods*. First, according to Johan Galtung,[1] peacebuilding is

> the process of creating self-supporting structures that 'remove' causes of wars and offers alternatives to war in situations where war might occur. Conflict resolution should be built into the structure and be present there as the reservoir for the system to draw upon, just as a healthy body has the ability to generate its own antibodies and does not need ad hoc administration of medicine.

Peacebuilding is coined in the context of post-conflict recovery work. Post-conflict recovery entails the promotion of reconciliation among the warring factions in a society or country (UNDP, 1994). It is about facilitating the restoration of durable peace which can also cover but is not limited to humanitarian relief, protection of human rights and assurance of security, including the establishment of healing process among all subgroups that have taken arms against one another (UNDP, 1994).

Peacebuilding is a complex endeavor and it is anchored in the participation of a variety of actors and institutions operating in diverse contexts and spheres. Among key actors are the citizenry; opinion leaders in communities/country; security and law enforcement agencies/officers; local and international aid workers; national policy and program managers; and victims and survivors of civil wars/social conflicts. In civil wars, quite a significant number of the actors may have been maimed and/or brutally traumatized. Indeed, the maimed/brutalized may be unwilling or unable to forget their pain and/or forgive the surviving perpetrators of various heinous atrocities in the course of civil wars or social conflicts. Healing process among human beings as we all well-nigh know takes time and only time tells if the efforts to reconcile victims and perpetrators of atrocities in post-conflict society can produce the desired results.

By and large, the complexity of the process in peacebuilding also brings into focus the role of key institutions in the exercise. The key and relevant institutions include a variety of civil society organizations; ethnic and religious bodies like the law enforcement agencies; state organs like ministries and departments; local/international aid workers; and international development partners. The international development partners always actively participate/facilitate peacebuilding efforts through huge financial contributions, logistical and institutional support in post-conflict societies.

Field-based research

The thrust of field-based research in peacebuilding however is about gathering baseline data/information on the role, activities of a wide range of significantly large number of individuals, subgroups, institutions and/or government and non-government organizations in various communities within distinct geographical sub-regions that have been engulfed in civil wars/social conflicts. The centerpiece of research work can be qualitative research methods utilizable for gathering enormous information on disparate groups, institutions and organizations spread over large geographic territory (ies) in post-conflict society. Variants qualitative methods such the participatory observational methodology, focus group discussion and key informant method can be very useful and helpful to gather data individuals, groups and in organizations (Salkind, 2010).

As we all well-nigh know, the study of human behaviors, institutions and organizations through sophisticated quantitative methods may not necessarily give a true and complete view of post-conflict societies that are eager to have peace. There is, as it is often argued, "stories behind stories." Quantification gives a side of the stories in figures which tend to mystify social reality. In contrast, the qualitative methods which are euphemistically referred to as "soft" methods, because they are bereft of quantification, often provide the "stories" behind statistical data on human beings, institutions and organizations that are generated through quantitative methods.[2]

Quantitative research methods

Quantitative research methods

> emphasize objective measurements and the statistical, mathematical or numerical analyses of data collected using computational techniques. The methods focus on "gathering numerical data and generating it across groups of peoples or institutions or organizations to explain a particular phenomenon.
>
> (University of Southern California, 2019)

Quantitative methods are always found to be far more appropriate than the qualitative ones for gathering data in contexts of several individuals/groups, institutions and organizations, in order to have a broad scope and range of data information. The information that is collected in such contexts will be easily amenable to quantification and they can also help us understand trends and forces, including how erstwhile warring subgroups in the societies can be reconciled to live in peace. As examples, one can quickly take a look at some of these reports, namely Alliance for Peace Building (2012); Filitz (2018); Benson (2018); World Peace Foundation (2014/2015) on diverse contexts. One other reason, numerical results obtained through quantitative method research help readers/researchers to can make easy sense of results a large body of data. Third, quantitative outcomes from surveys underscore the value of objectivity if efficiently undertaken by researchers. Finally, it is believed quantitative methods assure us of the replicability of the work/ outcomes in other contexts.

The variants for gathering data within the ambit of quantitative methods are: (a) *retrospective survey*; (b) *the longitudinal survey*; and (c) *cross-sectional survey*.

a Retrospective surveys focus on the collection of data that already exist or which are available. Examples are statistics on issues/persons/activities and so on; official documents that are kept in archives and so forth. Such records which are often kept by institutions or organizations are deemed to be or described as *secondary* data.
b Longitudinal surveys, also referred to as *time series* or *prospective*, however are designed to gather data on a continuous basis covering a specified time frame (e.g. a week, a month, a year or several years).
c The third type of surveys (i.e. cross-sectional) is designed to gather data at *a point in time*. For example, a researcher may wish to carry out a survey on attitudes of refugees to ascertain how well responsible authorities are responding to their plight or meeting basic needs at a point in time. Another may be mounted two years later or so. Each in this example is a cross-sectional survey.

Both the longitudinal and cross-sectional surveys are deemed to be focusing on the collection of primary data because they are being freshly collected. The strengths/weaknesses of each of (b) and (c) above variants will be discussed below.

On the whole, multiple competencies are required for quantitative research work. First, a researcher must have the capacity to conduct these sorts of surveys. Second, researchers must have a good grasp on techniques (i.e. sampling) for the collection of reliable and valid data from large study populations. Third, a mastery of data processing and analysis techniques is required. The researchers must have a good knowledge of statistics and how they can be interpreted. Fourth,, the ability to present information and write good reports are also essential.

Quantitative research methods – variants of survey

Retrospective survey method

Retrospective surveys are designed to gather secondary data. Secondary data consist of data that have already been collected or are already available in one form or another. They are sourced from documents, reports and statistical reports that are kept by institutions and/or organizations.

For example, one may wish to carry out research on repatriated refugees to their home country or of internally displaced persons within a country as part of peacebuilding efforts. Information on number/attributes of those repatriated and/or displaced in the civil war(s) or conflicts are likely to be in the records of the organizations that handle repatriation and/or resettlement of displaced persons. Their records may include but not limited to the following: (a) demographic attributes of these persons (i.e. age, gender, religious affiliations, marital status, ethnic or racial origin); health status; residency; budgets and expenditures in local/international currencies. These that are already available will be extracted and transcribed in numerical form for ease of entry into computer and subsequent analysis. A survey such as the foregoing example will be substantiated with numerical data to the efforts undertaken by the organizations that are responsible for repatriation and/or resettlement.

The aim of a retrospective survey in the specific case above is to take advantage of or rely on records that already exist in one form or another to be able to produce a report on work on the repatriated refugees/internally displaced persons in peacebuilding.

Data collection method

The first step for the researcher is to visit and seek permission of targeted institutions/organizations before commencement of the project. To begin with, it is advisable for prospective researchers to brief the head/officials of the targeted institutions/organizations and reassure them that their records will be treated with utmost confidence, and will also not be misused or abused.

The next step is for the researcher to inspect and review the records of the institutions/organizations that are available in order to have an idea of the range and magnitude of data to be accessed/collected. This will be followed by the design of instrument(s) to be used to extract information from the records. Needless to say the instruments that are to be designed are expected to be aligned to data that are being (to be) extracted from available records and documents.

Strength(s) and/or possible weakness(es)

Retrospective surveys of records have their strengths and weaknesses (MacMahan and Pugh, 1970). First, one of its advantages is that records to be used/

collected are already available and can be sourced quickly with the permission of their keepers/owners (i.e. the institutions and/or targeted organizations). Second, gathering data that are already available is much easier and cheaper than collecting afresh.

But there are a number of weaknesses in the use of records and documents. First, quality of records/documents varies from one organization to another. Some organizations keep records that are rich in information while others do not or may be tardy in record-keeping. The result is that researchers may confront discrepancies or missing information in the available data that are sourced from a variety of institutions/organizations that are targeted.

Second, and closely related to the above, is completeness or incompleteness of records probably due to poor record-keeping or deliberate effort by officials to manipulate records for administrative reasons. Third, records that are sourced may not be suitable for the issues that are vital to the researchers' interest (MacMahan & Pugh, 1970). For example, the demographic attributes of the repatriated refugees or internally displaced persons may not be robust enough for good analytical work. Lack of clarity or completeness of information or statistics will no doubt hinder rich analytical work and/or comparative analysis.

Cross-sectional survey

Cross-sectional surveys are designed to gather primary data. Primary data are data that are freshly gathered at a point in time. For example, a cross-sectional survey may be targeted at refugees who have been resettled in a particular locale. It is a one snapshot of refugees and/or in internally displaced persons in a locale at a particular point in time. A researcher may be interested in ascertaining the following concerning the repatriated or internally displaced persons: (i) what are their socio-economic attributes?; (ii) what challenges are they currently facing?; (iii) what are their health and financial challenges?; (iv) what have been their current vicissitudes?; and (v) how are the responsible authorities responding to them?

Data collection method

Such surveys will necessitate the design of an instrument (i.e. a questionnaire) that incorporates a wide range of questions on their demographic attributes (i.e. age, ethnic or racial origin, religion, occupation, marital status, income per annum; whether previously or currently employed; marital status; number of children if married; number of wives if male and number of co-wives if female in polygamous marriage;) including others concerning their conditions since repatriation and internally displaced; the challenges facing them; their coping mechanisms; and familial and social support system in place for them, if any.

The questions that are posed to the respondents may be *close-ended* or *open-ended*. Close-ended ones already have fixed possible responses to each of the questions from which a respondent is expected to opt for one that reflects his/her stand. The other type of questions is the open-ended ones to which respondents can freely give answers in their own words (Kerlinger, 1967).

However, an instrument (or a questionnaire) can be designed to contain an admixture of close- and open-ended questions or it can be entirely and completely open-ended (i.e. answers can be freely given in narratives to all questions in the questionnaire) or close-ended (i.e. questionnaire with range of suggestions given to each of the questions). Those that contain both the close- and open-ended questions are referred to as *semi-structured questionnaire*, while those that contain only close- or only open-ended questions are referred to as *structured instrument*.

A sample of the study population (i.e. the entire population of the total number of those repatriated or on internally displaced camp) will be ahead of data collection in the survey. Any sample drawn from the study population must fulfill two conditions – *adequacy* and *representativeness*. The former is the sample selected must be large enough to generalize findings while making sure that the sample is representative of the study population. A researcher can opt for any of the following sampling techniques to fulfill the foregoing conditions[3] (Baumgartner et al., 2002):

a random sampling;
b stratified sampling;
c quota sampling:
d aerial sampling; and
e purposive sampling.

By and large, each of the above sampling techniques has its merits and demerits that will not be discussed in the context of this chapter.[4] It is however important to note that the scientific worth of the outcomes of a survey will be in doubt if sample is not adequate and representative of study population (Phillips, 1966; Baumgartner et al., 2002).

Cross-sectional surveys which focus on large samples may require the assistance of assistants who will help in carrying surveys among the targeted in the sample. A Training Manual will be developed to guide assistants those to be identified and included in the sample; how to enter into homes or communities; and how to administer the questionnaire uniformly and successfully during data collection. Pre-field work exercise will be undertaken prior to full survey to ensure that all the foregoing is mastered by assistants. This will be followed by training of assistants on how to collect data and relate in an accustomed way to respondents.

In some context where there no common lingua franca, instruments for gathering data are translated into local languages for ease of communication between field assistants and respondents, and also to ensure that questions are posed properly and rightly and not due to the whims and caprices of the interviewers. Commonality in the way the questions in the questionnaire are asked or information gathered is central to the reliability and validity of the information gathered.

The collection of data in cross-sectional surveys may be undertaken by deploying assistants to administer questionnaire to respondents on face-to-face interview. This is referred to as the *interview method.* A researcher may adopt what is referred to as *mailed method* of collecting responses from respondents; in contexts, respondents are literate/speak same language and can read and write. Here respondents are expected to complete questionnaires by themselves and return their copy to the researcher on completion. While it is generally believed that mailed questionnaire approach on the face of this approach might appear to be cheaper than the interview method, this has proved to be not necessarily the case. Often a researcher using the mailed method is forced to expend more money because of low return from the targeted respondents in the sample. Efforts to address small numbers that are returned forces the researcher to telephone those who have not returned their money; make additional copies for further distribution and so forth. Meeting the conditions of adequacy and representativeness will be the primary reason for these efforts by the researcher.

Strength(s) and/or possible weakness(es)

The major advantage in a cross-sectional study is that researchers can design an instrument (or a questionnaire) to collect all the required data at the onset unlike in retrospective surveys. Consequently, data obtained in cross-sectional surveys can be remarkably rich and robust for analytical work. There is also ample opportunity for a researcher to collect richer demographic data on respondents than in retrospective studies. Good and reliable data on demographic attributes will make analysis of the interplay between independent and dependent variables worthwhile.

Even though cross-sectional surveys are laborious, they can nonetheless yield reliable and valid data if the sample is representative and adequate.

Data processing

The data from a cross-sectional survey are entered into the computer after they have been closely reviewed. All survey instruments will be serially numbered as follows depending on total number in sample: 0001–0004 (i.e. the last) while the data on other variables will be inserted in columns that follow identification number on the instrument. Data entry of a structured instrument is easier than that of the semi-structured. Therefore, steps will be taken to regroup the free responses to open-ended questions in unstructured and/or semi-structured one. A coding book (i.e. a replica of the entire instrument) is developed and deployed to be used to enter data into computer.

Analysis of data will follow data entry. Moreover, a preliminary review of the printout of deck of data that are entered will be reviewed to ascertain if there are errors during data entry into the computer. Errors are cleaned out. For example, an error may be entered as #2 for female in code book for Male when the code for Male is indeed #1.

The SPSS software[5] or any other software for data processing may be used to process data. Dummy tables are prepared ahead of analysis to assist researcher to identify variables to be focused upon during preliminary analysis of data, otherwise he/she might print out lots of tables which may not be needed to start with. Statistical results/findings are extracted from computer printouts, transcribed to dummy tables, interpreted and narratives appended to the dummy tables. Finally, narratives on drafts results emanating from dummy tables are presented and used to produce the final report of survey. Data analysis is further discussed below.

Longitudinal survey

The third variant is longitudinal survey, which is also referred to as *time series* or *prospective survey*. It is designed to gather primary data over short or a longer period of time. As an example, survey aimed at gathering data starting at a point in time and covering a week, a month, a year or years. Data gathering in such a survey is a continuous exercise and could begin, for example, from January 1, 2018, to December 31, 2020 (i.e. three years) or for many more years.

Data collection method

An instrument which may be close- or open-ended and which takes into account the general/specific objectives of survey is designed to gather data from the respondents.

Before focusing on longitudinal surveys in their true sense as described above, one can also undertake another form of longitudinal survey but this time as a retrospective one. This is about taking a backward look or study of available records. This form is not usually refereed as longitudinal in the literature, but one could also specify it as some sort of one.

An example is a survey of past records, documents and databases over many years in the study of social unrests, beginning from several years in the past and culminating in a civil war that eventually took place in a country. This sort of study will entail review of reports in a country's national newspapers, government documents and reports on unrests, reports of commissions of enquiries into past social unrests, all of which are likely to have been stored in national archives. Such sources will help the researcher gather the following:

a Number and dates of civil unrests;
b Number of perpetrators and/or victims;
c Frequencies of civil unrests per year etc.;

d Sub-regions or local government areas where they took place;
e Factors accounting for unrests;
f Number of casualties and deaths associated with unrest;
g Response of national and local authorities;
h Location of unrest (may be urban or rural) and so on and so forth.

The work can be complemented with in-depth interviews (a qualitative method) with survivors, victims of atrocities, government officials still alive and community leaders still alive who fill gaps by furnishing additional information and so on.

This is a kind of retrospective survey of going back in time in history to collect data that can be quantified. It should be noted that some of the weaknesses previously highlighted concerning recognized retrospective surveys would also apply here. However of concern here are surveys that focus on continuous collection of data at a point in time over a specified period of time as described above (i.e. a longitudinal survey from a prospective standpoint).

Data processing

The next stage after data gathering is data processing, which will be a continuous exercise because of the huge amount of data collected over time. It is vital to put in place strategies for handling a huge amount of data from the onset depending on the scope/magnitude of data. The procedure for handling data obtained in cross-sectional and/or retrospective surveys will also apply here. This involves the development of a code book for data entry, having taken all necessary steps to handle open-ended questions/responses, cleaning of data to ease the work of data entry.

Strength(s) and possible weakness(es)

Longitudinal surveys can be time consuming, energy sapping and also expensive because data collection is a continuous exercise over a period of time – a month, a year, two years or several years. One of the most important studies on maternal morbidity in low-income developing countries was a 20-year prospective survey carried out in Zaria, Northern Nigeria, by Harrison (1985), and another was a 21st-century prospective survey on hazards of smoking and benefits of stopping in the United Kingdom reported by Pirie et al. (2013). It means lots of work with several persons possibly and collectively assisting in data collection, monitoring of data gathering; management of data; data entry; and so on. The capacity to process huge amounts of data therefore becomes paramount. Longitudinal surveys yield important and huge amounts of data from which researchers can produce and do definitive work.

By and large, researchers often shy away from longitudinal surveys because of the cost and time implications. The method is more taxing than a cross-sectional or retrospective survey.

Surveys among special subgroups in field-based peacebuilding research

It is necessary to focus on special subgroups that are likely to be targeted in peacebuilding efforts. These are the traumatized women, girl-children or boy-children as well as physically challenged among refugees and internally displaced persons. These subgroups cannot be treated like other persons in the study sample because care should be taken not to cause them further pain.

Researchers should bear the above in mind because one of the important ethical principles in conducting research on human subjects is not to cause them harm. Reports that emanate from surveys that violate this ethical principle among special subgroups will be flawed and are not likely to be well received in the scientific community. Further discussion on ethics in research on human subjects and/or in the context of the use/access to records and documents belonging to organizations will be discussed below.

A variety of other techniques is used to mitigate pain or harm in the course of surveys. A researcher can opt to use projective test techniques or vignettes to gather data from these special subgroups (Kerlinger, 1967). Projective tests are based on photographs and diagrams depicting sexual abuse or violence to which respondents that are targeted can react and which enables researchers to gauge attitudes/experiences of target subjects rather than resorting to asking direct questions and/or probing respondents on such sensitive events in their lives. Vignettes however are a collection of descriptions of activities or behavior patterns provided in the third-party narratives to which respondents can react to in the course of interviews. Respondents' responses can be used to gauge their pain and personal experiences. The techniques are useful in cross-sectional and longitudinal surveys that target vulnerable women, men, young girls, children and the physically challenged ones, and have been widely used in surveys on mental health issues (Erinosho & Ayonrinde, 1978).

Data analysis in quantitative research methods

Having described the various quantitative research methods in the context of peacebuilding research with a few illustrative examples, it is time to focus on data analysis. Collection of data using quantitative methods is one thing, but how to process the data that are collected is equally important. We previously discussed aspects of how data are coded and entered into computer in preparation of analysis and there is no need to repeat the steps here again. The aim in the following discussion is to underscore what is needed to do good analysis in surveys that are previously outlined.

Knowledge of statistics

First, a good analysis of quantitative survey data will well-nigh depend on a good grasp of statistics. The researcher who is acquainted with his/her research on aspects of peacebuilding must *not depend* entirely on programmers and statisticians to do data analysis, but can seek their assistance if necessary. He/she is the person most acquainted with the objectives of his/her survey, and he/she can figure out the key variables and statistics that are most useful and appropriate to interpret results and help readers to grasp the essence of the work. Sometimes, a good research work may be poorly presented to readers. Besides, poorly presented results/statistical analyses can provoke lots of comments and questions which could otherwise be avoided. Therefore entrusting data analysis entirely to programmers/statisticians who may not grasp the objectives of surveys may lead to one churning out results that will be meaningless.

The above notwithstanding, the first step is for the researcher to test run data that are entered into the computer in order to ascertain if errors may have occurred during data entry. For example, a variable like gender which ought to have been coded as follows: #1 for Male and #2 for Female may turn out to have other values like 3 or 4, which are out of place in the code book when data are test run. This then indicates that there were errors during data entry. Such entry errors will force the researcher to eliminate these errors only after going back to cross-check raw data in the instrument vis-à-vis the printout. This is important because entry errors are missing values that will affect results during data analysis, unless otherwise indicated as such during analysis. Data processing can begin in earnest once the foregoing has been accomplished.

The next step is for the researcher to produce the computer printout of frequency distribution of all variables. The printout of frequency distribution provides an idea of the number of responses per each of the variables in the data deck. It will also indicate mean, median and standard deviation for each of the variables. The mean of an ordinal variable enables the researcher to determine the cut-off point for recoding variables like age into young and old or education into low education and high education, respectively, for meaningful analysis.

The fourth step is to focus on demographic variables (e.g. age, gender, education, religion) which are deemed to be independent variables vis-à-vis the dependent ones. This step will produce a number of cross tabulations tables for close review/analysis. Cross tabulation tables provide a researcher ample opportunity to do further analysis.

One is expected to recode variables like age or education to make sense of tables that are printed out. As an example, the variable age which would have been spread out in the raw entry into computer is as follows:

#1. 10–12 years;
#2. 21–30 years;
#3. 31–40 years;
#4. 41–50 years and
#5. Above 50 years
0. No information

The above variable (age) can be recoded into young and old, using mean score extracted from frequency table as the cut-off point for sub-dividing the sample/respondents. The Chi-square test result, for association between *age against a dependent* variable if the latter is also recoded into two values in a two by two table, will make sense if the tables are on cross tabulations of 3 by 5 or 2 by 6 tables and so on.

Be that as it may, frequencies and cross tabulations merely scratch the surface in data analysis. Cross tabulation tables often do not take care of intervening variables which will undermine the so-called statistically significant results even in 2 by 2 tables. It will be desirable to undertake further analysis to take care of possible intervening variables which will necessitate more rigorous analytical work, using inferential statistics like regression, step-wise regression, analysis of variance and path analysis.

Analysis on this scope requires knowledge of how to use nominal variables like Male or Female; occupational categories; or religious affiliations and ordinal ones like age, education and more. The nominal variables are expected to be converted into dummy variables before they can be used in more sophisticated analytical work, otherwise the researcher would end up churning out meaningless findings. Here a programmer or statistician can help in the conversion of nominal variables into useable ones in the context of inferential statistics like step-wise regression analysis.

One final note is that the more rigorous data analysis, the better and the more robust presentation of results and conclusions arising from a study.

Ethics in quantitative research methods

Consideration of ethics in research involving human beings, animals and organizations has always been of vital importance. Researchers are expected to outline in the protocols of their survey how they will comply with ethics in research focusing on humans/organizations in the course of peacebuilding research work. Protocols will be reviewed and only approved by the institution's ethics review or the international board (i.e. if grants for project are being sourced from outside agencies) if ethics will be complied with in the course of survey/study.

A number of ethical principles that guide surveys are outlined by Nuffield Council on Bioethics (2005, 2007) and Erinosho (2008):

- a Confidentiality of information and records
- b Consent seeking
- c Respect for respondents
- d Avoidance of harm to targeted respondents
- e Recognition of information sharing and feedback to respondents
- f Integrity of data

Confidentiality

The confidentiality of documents and personal and official records that are sourced in the course of retrospective surveys must be assured by the researching persons or bodies. The data sourced from records and documents must be protected. For instance, personal information sourced on respondents must be carefully handled and protected. Instead of indicating the names of the respondents on the interview schedule, a number will be assigned while names will be kept in a separate register to safeguard confidentiality of information on personal details. Many ethical review boards now demand formal narratives from researchers on how confidentiality will be assured in the course of their study. Violation of the confidentiality principle is considered a serious lapse in survey research work

Consent seeking

The second is consent seeking. The respondents who are targeted in the course of surveys must be briefed before interviews are conducted. The permission of each respondent must be sought prior to interviews and respondents are free to agree to be interviewed or refuse at the onset of an interview. Their decision must be respected at all times. Consent also covers the process of interviewing respondents. This means that the right of respondents not to continue to be interviewed after an interview has commenced should also be respected.

Ethical or institutional review boards of funding agencies and universities worldwide now insist that consent be given in writing or by the signature of respondent on the consent form or by thumb printing consent form before interview begins. In some contexts, consent may be given orally (Adebamowo et al., 2008). This requirement of consent in writing seems to pose a serious challenge to researchers working in non-literate communities in Africa where people are not inclined to give consent in writing or by thumb printing for a variety of reasons; hence the demand for oral consent. There is indeed no way of getting around the principle of consent seeking before interview.

Consent seeking in the case of mailed questionnaire differs from the above. Here the researcher is expected to attach a consent form to the questionnaire or insert a preamble on first page of his/her questionnaire/instrument the need for consent which respondents will sign.

Respect for respondents

Closely linked to the above is respect for respondents in the conduct of surveys. It is absolutely important to show deference to all respondents and most especially to those who might have been traumatized as a result of civil wars/ social unrests. Here we are talking of internally displaced persons or repatriated refugees many who are likely to be trapped in unsavory conditions/

circumstances in a refugee camp. Approaching/interviewing them in the course of data/collection ought to be handled with considerable finesse in order not to arouse past/current painful memories or experiences.

One can begin by gaining entry into such subgroups (i.e. repatriated/internally displaced persons) through the officials in charge of camp, who in part will introduce researcher and also facilitate information gathering in a skillful manner. The main reason for these steps is not to pounce on respondents.

Avoidance of harm to respondents

The fourth of the ethical principles indicated above is about avoidance of causing harm to respondents. Every research work is not expected to bring harm to respondents in whatever way or form. The questions that are posed to respondents in the course of conducting surveys using instruments/questionnaires should never arouse anger or trauma or pain in the course of data collection during survey. As such, all questions that are likely to do that should have been removed during pilot study.

The targets in surveys of those who have been part of civil wars and/or devastating social conflicts are most likely to carry scars of wars and it will not be appropriate to add more to their pain in the course of interviews during surveys.

Some segments of the sample like girl-children or women who might have been sexually and/or physically abused in the course of wars and social unrests will also require special treatment such as not asking intrusive questions. Surveys among such will require the use of psychometric tests and/or vignettes. These strategies have been elaborated for surveys among similar vulnerable subgroups.

Special explanatory notes on how to handle such subgroups in the course of cross-sectional and longitudinal survey must be provided to ethical review boards, otherwise protocols for surveys may not be approved. Ethical review boards are mindful of the harm that intrusive questions can cause to the vulnerable and the already emotionally traumatized.

Information sharing and feedback

Information Sharing/Feedback is also an important ethical principle that should be safeguarded. Researchers are continually targeting victims and participants in civil wars/social upheavals. Studies upon studies by various interested individuals and organizations are commissioned and undertaken from time to time among these various subgroups and there is ample reason to believe that they are being over-studied.

Oftentimes researchers move in to conduct their surveys and they disappear without feedback to those targeted. Respondents are overburdened and fatigued as a result of repeated efforts by "outsiders" to study their conditions and challenges. As such, the question that is often asked by the targeted respondents is: what is in it for the myriads of studies on us? Or how can you through your study ameliorate our plight/conditions?

This has led to the idea of giving something back to the over-researched targets. While researchers may not be able to give respondents tangible rewards like money they can at least share the outcomes of surveys with all respondents, organizations, including the displaced and traumatized persons living in camps, for their cooperation.

Today many researchers are enjoined to go back to those who have been studied to share the outcomes of their surveys with them as well as take questions and/or clarify the outcomes of their surveys. An additional advantage of feedback is that it helps researchers to revisit sites, revisit their results in case these might have been misconstrued/misinterpreted in the course of data analysis and discussion of findings.

Integrity of data

Finally, there is the principle of the integrity of data. This is a sacred principle that should guide project implementation in order for the process and outcomes to earn respect in the scientific community. The samples selected must be adequate and representative and reasons given if these conditions are not met. Attempts to mask violation of ethics in the course of survey as well as plagiarism should be avoided. Possible bias or limitations in the results of survey should be clearly outlined to readers of reports.

The process underlying fieldwork should be carefully thought out and free of bias of any kind. Field assistants (i.e. in case their services are utilized during survey research work) should be trained along with oversight supervision of their work so that the data that are gathered are reliable and valid. Instruments should be properly administered. Finally, efforts must be made to provide robust analysis and interpretation of data so that the issues are clear and no attempt is made to manipulate data/results for self-serving reasons. These are critical issues in the implementation of surveys that are targeted at respondents in field-based peacebuilding research.

Reporting and analyzing research findings

Surveys are incomplete without good reports on their outcomes. Consequently, report writing is a vital component of quantitative research methods.

The results that are printed out from computer will be sorted into separate files that are earmarked for each section of survey reports. Narratives are added to tables and with clear indication if results that are extracted are statistically significant on not. All of these will ease narrative of results and discussion section by section. The aim is to constantly amplify clearly the relationship between the independent and dependent variables which are contained in the specific objectives of surveys. Findings are chronologically discussed to bring focus to survey report.

Ample information should also be provided to show how results are arrived at, especially in the context of inferential statistical analyses. For instance, there will a need to highlight how dummy variables were converted in the course of step-wise regression. Discussion will follow analysis and interpretation of results by each of the objectives. Narratives on tables will indicate level of significance of results. Indeed, the organization, presentation as well as full discussion of results chapter by chapter is key to good reports. Results must be interrogated against the backdrop of the following:

a Background literature to cover studies from global to local to cultural context of survey; other unpublished reports on subject matter of survey; possible explanations for discrepancies between the writer's analysis and others.
b Context of the survey covering historical antecedents and background to civil war or social unrests; factors accounting for social upheavals; consequences from already existing documents, reports and so on.
c Overview of methodology of study which will cover: (i) study population; (ii) sample and sampling; (iii) data collection method (including description of instruments used); (iv) shorts overview of how ethics was addressed in survey; (v) data analysis (to include various types of analysis whether descriptive and/or inferential statistics); (vi) challenges confronted during survey and how they were surmounted; including (vii) limitations of the study/survey.
d Findings of survey which may be spread into one or more chapter(s) depending on the scope/magnitude of data from a survey.
e Finally, a good report cannot but be concluded without a chapter/section on recommendations for policy and program development. Such recommendations are usually for the careful consideration of funders of research; interested persons and bodies/organizations; or responsible national or local authorities. Recommendations must be based on results that emanate from a survey.
f Recommendations must be amply justified based on results otherwise they might be dismissed outright. On the whole, the recommendations emanating from survey should not be "woolly" or unfounded or simplistic or not implementable.

Conclusion

Quantitative methods can be used in a variety of ways as outlined in this chapter. But all the variants of quantitative research methods can be used in a single study. It may well-nigh be appropriate to conduct a retrospective survey on aspects of peacebuilding and also undertake a cross-sectional one in the same study and/or also embark on a longitudinal survey in the same context or population. The decision to embark on such a huge endeavor will depend on the scope of the general and specific objectives of the study, the timeline for study and the resources that are available to researchers.

Other research methods can also be used in the context in which the primary/principal method for data gathering is anchored on surveys as outlined above. For instance, variants of qualitative methods can/may be used alongside the quantitative. A combination of variants of the quantitative and qualitative research methods will further enrich the outcomes to be produced from such an endeavor.

In concluding, one can therefore envisage the use of *mixed research methods* in field-based peacebuilding research. The ultimate goal of any good or worthwhile research is to gather as much evidence as possible with particular reference to the general/specific objectives that are outlined in the approved protocol of the study. In other words, the aims of any study include the production of evidence-based and worthwhile reports that are devoid of unsavory criticisms in the scientific community.

Notes

1 See Selected Definitions of Peace building Alliance for Peace Building, August 12, 2013; Type Peace Building News: Tags: Conflict Assessment, Technology and Innovations; and also see J. Galtung, *Three Approaches to Peace: Peace Keeping, Peace Making, and Peace Building: In Peace, War and Defense: Essays in Peace Research* (1976), Vol. II, ed. J Galtung, Copenhagen Ejlers.

2 Sociologists and others euphemistically suggest that data from quantitative research methods are "hard" while those of qualitative are "soft." This view arose out of the debate on positivism. For a long time, the debate is whether the study of human behaviors in groups and/or organizations, communities and institutions can really be deemed scientific or humanistic discipline. Human behaviors can be subjected to exactness, as it is in the study of atoms, molecules and so forth in science, that is, whether we can truly have a science of society (i.e., quantifying outcomes of our research work as in physical and biological sciences due to the shifty nature of humans). Indeed many others are inclined to argue that disciplines like sociology will ever remain in the humanities because of the likelihood of subjectivity due to various factors that cannot be fully discussed.

3 Oftentimes, specialists in sampling techniques or demographers may be consulted to help in the selection and determine adequate sample in a given study population as well to ensure representativeness of a sample

4 For more technical details on sampling procedure, see Kish, L. (1965) *Survey Sampling*, New York: John Wiley & Sons.

5 There are a number of software in use like SPSS or EPInfo. There is a simple SPSS Version 16.0 for data analysis; see L.A. Kirkpatrick and B.C. Feeny, *A Simple Guide to SPSS for Version 16.0*, Belmont, CA: Wadsworth Cengage Learning, 94002-3098, 2009.

References

Adebamowo, C.A., Mafe, M.A., Yakubu, A.A., Adekeye, J.M., & Jiya, J.Y. (2008) "Developing Ethical Oversight of Research in Developing Countries: Case Study of Nigeria," in Erinosho, O.A. (Ed.) *Ethics for Public Health Research in Africa*, Abuja: Social Science Academy of Nigeria, pp. 7–15.

Alliance for Peace Building. (2012) *Peace Building 2.0 Mapping the Boundaries of Expanding field*, https://allianceforpeacebuilding.org/wp-content/uploads/2013/08/AfP-Mapping-Report_online_FINAL.pdf

Baumgartner, T. A., Strong, C. H., & Hensley, L. D. (2002). Conducting and Reading Research in Health and Human Performance (3rd ed.). New York: McGraw-Hill.

Benson, J. (2018) *The Geography of Violence against Civilians: Implication for Peace Enforcement*, An OEF Research Division Paper, OER Research, www.oefresearch.org.

Erinosho, O.A. & Ayonrinde, A. (1978) A Comparative Study of Opinion and Knowledge about Mental Illness in Different Societies", *Psychiatry – Journal of Interpersonal Processes*, Vol., 41, No. 4, pp. 403–410.

Erinosho, O.A. (2008) *Ethical Issues and Research Code in Social Science Methodology in the Context of Africa, in Ethics for Public Health Research in Africa*, Abuja: Social Science Academy of Nigeria, pp. 37–42.

Filitz, J. (2018) The Collapse of Mozambique's 2013 Peace Agreement and Why It Matters, An OER Research Discussion Paper, OEF Research – Informing Change for Peace, United States Marine Corp https://oefresearch.org/sites/default/files/documents/publications/mozambiqueDisscusionPaper.pdf

Harrison, K.A. (1985) "Child Bearing, Health, and Social Priorities: A Survey of 22,774 Consecutive Births in Zaria, Northern Nigeria", *British Journal of Obstetrics and Gynaecology*, Vol. 92, Supplement 5, pp. 1–119.

Kerlinger, Fred N. (1967) *Foundations of Behavioral Research: Educational and Psychological Inquiry*, London: Holt, Rinehart and Winston, Inc.

Kish, L. (1965) *Survey Sampling*, New York: John Wiley & Sons.

MacMahan, B. & Pugh, T.F. (1970) *Epidemiology – Principles and Methods*, Boston: Little, Brown and Company.

Nuffield Council on Bioethics. (2005) *The Ethics of Research Related to Healthcare in Developing Countries*, London: Nuffield Council on Bioethics.

Nuffield Council on Bioethics. (2007) *Public Health: Ethical Issues*, London: Nuffield Council on Bioethics.

Phillips, Bernard S. (1966) *Social Research: Strategy and Tactics*, New York: Macmillan.

Pirie, Kristin, Peto, Richard, Reeves, Gillian K, Green, Jan, and Beral, Valerie (2013), "The 21st Century Hazards of Smoking and Benefits of Stopping: A Prospective Study of One Million Women in the UK. *The Lancet*, Vol 381, Issue 9861, pp. 133–141. doi:10.1016/S0140-6736(12)61720-6

Salkind, N.J. (2010) "Field Study", in *Encyclopedia of Research Design*, https://dx.doiorg/10.1435/97814129611258.n152, Sage Research Methods.

UNDP (United Nations Development Programme). (1994) *Human Development Programme Report*, New York: Oxford University Press for UNDP.

University of Southern California. (2019) "Organizing Your Social Sciences Research Paper: Quantitative Methods", https//libguides.usc.edu/writing guide/quantitative.

World Peace Foundation. (2014/2015) "Civil War, Mass Atrocity Ending of 7 August 2014", httpa://sites.tufts.edu/atrocityendings//2015/08/07/Mozambique-civilwar/

Case studies

7 Understanding complex new wars in Africa

The example of the Sahel

Amy Niang

Introduction

Since 2012, the Sahel has been plunged into a permanent state of warfare and turned into a transnational battleground as well as a field of experimentation for various modes of military interventions. As the conflict enters its ninth year, reasons for optimism are scarce. Isolated attacks on civilians and military installations have become more deadly, more widespread, but also more diffuse and less centered on conventional political claims. The proliferation of armed and militant groups in the Sahel since the onset of the conflict seems to confirm a widely held view that the problem of the region is one of governance. The Sahel is a region where the state has little, if any, control over vast territories, uneven processes and diverse peoples. It offers the image of a fractured geographical space that harbors various types of extremism, ranging from the religious to the ethnic. Indeed, what started as a localized, anti-state, secessionist rebellion in northern Mali almost a decade ago has now become a crisis that has engulfed the central and southern parts of Mali and the entire Sahel.*

Despite the presence of the United Nations Multidimensional Integrated Stabilization Mission in Mali (MINUSMA) and the French-led anti-terrorist Barkhane force as well as the joint G5 forces, attacks by armed groups have become more frequent and deadlier. In 2019, these groups were responsible for over 1,500 civilian deaths according to the United Nations. They have also increasingly directed their attacks at non-combatting populations in Mali, Niger and Burkina Faso. The situation has spurned volatility, militarization and deterioration of security in the region in the past few years.

The complexity of the situation in the Sahel has challenged the meaning, purpose and efficacy of certain concepts. In particular, it has called into question the notion of the War on Terror (WoT) and its associated ideological baggage as well as its pervasive and disruptive policy apparatus. How should we understand the mobilization of WoT's belligerent terminology by both national and international actors to mitigate insecurity and to restore "stability?" How useful are discourses and policies of counterinsurgency and containment that have stigmatized societies in certain regions engulfed in the global campaign, especially given that these policies are reminiscent of colonial rationalities and their specific imaginary of spaces of security linked to a capacity to control subject-bodies? These questions call for serious reflection on the Sahel as a space of epistemological and empirical experiments. The experimental mode affects virtually every aspect of the region's political, economic, cultural and spiritual configurations.

This chapter aims to do four things. First, it provides an overview of the developments of the conflict in the Mali and the Sahel since 2012. It shows how the resurgence of violence since 2015, in particular in central and southern Mali—regions that had thus far been spared from widespread violence—is an outcome of the disintegration of structures of livelihood in a general context of mistrust. Second, it analyzes security and neglected perspectives on (in)security, especially from the perspective of the recipients of security interventions. Third, it shows how the Sahelian conflict is an example of a complex new war (CNW) that is multifaceted and entangled in broader historical, regional and global processes that frame local dynamics. Finally, it reflects on the implications for studying CNWs in Africa. It is no longer possible to study the conflict in Mali and in the Sahel the way conflicts in the Third World were once studied, that is, as "ethnic wars" in savage, illiberal places confronted with the imperative of change and liberal progress in a postcolonial world (Sogge 2011). Rather, the Sahelian conflict is part of global transnational shifts articulated by geopolitical tussle for control over resources and transit routes for the transportation of resources, people's mobility and "corridors of security". The underlining concern has much to do with both epistemology and methodology. This chapter suggests that research endeavors on CNWs should entail the use of innovative tools as much as a rethinking of the varying acuity of concepts and methodologies used.

The context of instability in Mali and the Sahel region

Despite various initiatives, namely the Algiers talks and the Bamako Agreements of 2015 to end the recent conflict, a spate of attacks, including that of the Radisson Blu Hotel in the same year, have stalled the peace process in the Sahelian region.[1] Mali continues to be a divided country under the uneven control of government security forces on one hand and a number of autonomous groups on the other hand, some of which have federated and fractured over time. Some of these groups have antagonistic or competing goals and have therefore clashed on several occasions. The Malian armed forces or FAMA are predominantly positioned along the Niger River Bend and in Tessalit, alongside French soldiers; smaller military units are also positioned in strategic areas such as Kidal, Aguelhok or Taoudéni.

The landscape of violence in central and southern Mali

The context

The current situation in Mali and the Sahel more generally was triggered by the 2012 Tuareg nationalist uprising that surprised and destabilized the Malian armed forces. The Tuareg nationalist movement was in turn defeated by Al Qaeda allied to other jihadi groups. In January 2013, France intervened to curtail the growing conflict upon formal request by the Malian President Diokounda Traore. By then, the jihadi groups had effectively occupied Kidal, Gao and Timbuktu and were pushing closer to southern Mali. Operation Barkhane followed operation Serval, led by France and supported by the African-led International Support Mission to Mali (MISMA), a 4,700 strong deployment predominantly made of Special Forces. On 25 April 2013, through UNSC Resolution. 2100, MINUSMA was established and it was mandated to stabilize and support the political process in Mali. MINUSMA temporarily pushed Al Qaeda, the Movement for Unity and Jihad in West Africa (MUJWA) and other Jihadi groups out of the cities they had occupied. These groups have since redeployed in the north and have in fact been expanding in the central and southern regions.

The Tuareg and jihadi insurgencies have turned northern Mali into a contested and unstable region, thus exacerbating existing social and religious divisions. The insurgencies have produced two trends that have major implications for the sustainability of the post-colonial Malian state and for establishing durable peace in the Sahelian region. First, the Tuareg demand for statehood has radicalized the autonomist and separatist movement in the Sahel. Second, anti-Tuareg sentiment coupled with degrees of resentment towards the Malian state have polarized inter-communal tensions. In central and southern Mali, militant groups and self-defense militias proliferated within a brief period in the absence of the state because they could exploit local conflicts to their advantage. Local conflicts facilitated and accelerated the recruitment and the constitution of ethnically homogenous formations.

The disproportionate focus on secessionist tensions in the north has meant that sources of instability in central and southern Mali have largely been neglected until recent times. Central Mali has long been a recipient of violence from the north, which led to the upsurge of banditry and availability of weapons since 2011 as well as increased cattle theft. Since 2012 and more intensely since 2015, there have been several armed attacks targeting communities in central Mali. In the 2012 crisis, in fact, the National Movement for the Liberation of Azawad (MNLA) and MUJWA occupied parts of central Mali, namely the Mopti area, and they sought to recruit amongst the youth there. In Macina, the Nterere first joined the MNLA and later MUJWA either by conviction or opportunism, or perhaps to find protection under an armed group and in the absence of government protection.[2] In the play of alliances that prevailed in the 2012 period, the salience of ethnicity was already apparent. For instance, Fulanis from the Macina joined MUJWA training camps in Gao, with the view of combatting the MNLA and the Tamasheq that make up its majority. The same move however was also an attempt for Fulani herders to gain an upper hand over Dogon sedentary famers in the longstanding conflict over access to pastures and fertile land.

The tensions in the central part of Mali have been concentrated on two main areas: the Macina, between Mopti and Segou, which is the political and historical region of the Fulani, and Hayré in the department of Douentza in the north-east of Mopti. These tensions preceded the 2012 crisis and became amplified in the post-2012 period. Amadou Kouffa's group has been dubbed the Ansar e-deen Macina, therefore katiba or battalion of Iyad ag Ghali's group. Kouffa's group has been targeting civil servants and government representatives in sporadic attacks since 2015, forcing the latter to flee to Bamako. Kouffa is said to have joined the Tabligh Jama'at between 2009 and 2011; the latter is a preaching movement born in Pakistan. It is also known in the Sahel under the name Dawa. At any rate, previously existing mechanisms of cohabitation in the Macina have been put under severe stress on account of the further deterioration of the relationship between the Fulani and their neighbors in central and northern Mali. A direct result of such deterioration has been resentment and hostility directed at the Fulani. The transformation of the image of the latter from herders into mujahideen was facilitated by the FLM activism—which was reactivated through the instrumentalizing of various stereotypes.

Islamic radicalism and ethnicity

Two trends—amongst many others—capture the complexity of the "Sahelian crisis". First, the implications of Tuareg demand for statehood and of the Tuareg question more generally for political stability in the Sahel. Second, the rise of armed, particularly self-defense, groups in northern and central Mali lays bare the limits of militarism and the securitization approach to local problems. The two phenomena are closely related to two parallel processes, namely religious radicalization and the polarization of inter-communal differences. These however cannot be seen as causes, but rather consequences, if not manifestations, of a conflict that has roots in historical and social configurations. In fact, according to Raineri (2019), islamized radicalism is now competing with ethnicity as a driving force for identity-based mobilization.

The issue of radicalization is a reality, but it should not be reduced to an account of the expansion of jihadist philosophy in historically "moderate" spaces of Islam. Sahelian radicalizations have specific contexts, and these contexts vary across countries and across regions within the same country. Radicalism had become prominent and widespread but it has to be understood in its various local articulations. Minorities have understood that in order to be invited to peace/political discussions—in fact in order to be listened to at all—they have to take up arms. Armed uprising thus becomes a shortcut to the negotiation table and insurgency an effective mode of engagement with the state. The political context is thus the specific appeal of armed rebellion following a historical pattern whereby only armed groups are seen as valid interlocutors in conflict resolution and settlement processes. While many analysts are quick to see jihadism and religious radicalism as *the* problem in these intersecting waves of violence, Islamism is not the root of the psychosis. Neither is ethnicity or cultural difference. The violence is rooted in the disintegration of the structures of livelihood and the structures of peaceful coexistence among various communities. The question of land is primordial here. The lack of legal clarity over land rights for nomadic communities and their settler counterparts has been one of the main contentious factors.

Even though land and land rights constitute overarching issues in recent developments, the precise triggers and motivations for the upsurge of violence in different localities are not easy to pin down. The nature of alliances amongst various groups, old and new, depends on political contingencies, opportunism and a complex set of calculations. For instance, in 2012, in the Méma and the Farimaké, Ntéréré first joined forces with the MNLA before switching alliances to join MUJWA when the latter started defeating the former. Equally, the Al-Mulathameen ("the Masked") of Bel Moktar left Al-Qaeda in the Islamic Maghreb (AQMI) to form a splinter group and later on joined the Movement for Oneness and Jihad in West Africa against AQMI and the MNLA. At any rate, the general climate of insecurity, chronic instability and deep distrust produces not only a volatile situation, but also the most unusual alliances which constantly shift centers and targets of violence.

Despite the fluid nature of alliances, the deep discontent of minorities in the north and the center against Bamako remains one of the most enduring features of Malian politics. In parallel, the growing impatience among young people and former dependents with regards to the slow transformation of old hierarchies into fairer structures feeds into historical resentment. The relative authority conferred by membership into an extremist group is an opportunity to turn the hierarchical table; hierarchies emerge from political elitism, economic wealth, social privileges or ethnic belonging. The political economy of violence has reshuffled the social order within various cultural groups. The possession of weapons is a solid argument for former social dependents in their contestation of both traditional local aristocracies and urban elites. Internal to many groups such as the Fulani and the Tuareg, there is an ongoing struggle between old and new generations, dominant categories and social dependents. Instantiations of local rationalities include the rebellion of Fulani pastoralists against dioros or the traditional leadership structure, Imghad Tuareg's contestation of Ifogha aristocratic leadership, the rise of former social dependents and slaves against hierarchical peers, young people joining radical groups as a way of contesting elders' authority and so forth.

The political context is therefore one of an "offer" of protection for populations exposed to volatile security situations, often by the very same protagonists involved in insecurity-generating activities. The label of "jihadist" often becomes appropriated by interest groups such as the Fulanis themselves. The Fulani rebellion and acts of resistance have taken a "jihadist turn" as a result of a combination of opportunism, political strategy and a desire to amplify the scope of relevance for their cause.

The trajectory of the various groups involved in recent armed conflicts and the nature of attacks have something to tell us about the sociology of violence and the (re)sources of instability in the Sahel. If current configurations are a result of the recent 2012 crisis, the more profound root causes of the crisis are to be found in a longer history, specifically the uneven history of postcolonial state-building, the absence of an adequate mode of engagement between state and society in Mali and structural, economic and political disparities between North and South.

The Malian state and its discontents

The Malian state's attitude towards some of the internal struggles mentioned above has been ambivalent in the sense that on one hand, it has advocated the abolition of all forms of hierarchies based on "race", social status and other "feudal" practices. On the other hand, the state has relied on aristocratic leaders in a bid to maintain patronage politics. In the context of the post-2012 crisis in particular, armed groups have emerged in the vacuum left by radical groups that had occupied central Mali during the 2012–2013 period. If militants have been partly neutralized or dislodged by Operation Serval, the return of a weakened Malian army and the resumption of persecution of ethnic minorities, specifically Fulanis and Tamasheq, have driven these groups into various self-defense groups.

The state's response to violence has been security-oriented for the most part. Decades of sporadic insurgency have demonstrated that repressive security policies have many limits. These are not only counterproductive; they also divert resources from education, health and other social sectors while bloating defense budgets. Where political solutions were sought, they were at best ambiguous and were in no way informed by the kind of long-term engagement contained in the Algiers peace talks; political actors in central Mali were, for instance, not included in the talks. At any rate, national reconciliation has never been a well-structured or enduring process in Mali. The successive accords of Tamanrasset (1991), the National Pact (1992) or the Algiers Accord (2006) were never fully implemented. Equally, the Bamako Agreement of June 2015 is yet to be implemented to the satisfaction of parties involved.

The rejection of the state is nothing new for it has more often than not been encountered and experienced as a repressive machine. In fact, beyond the terrorist dimension, some of the ongoing contestations in the Sahel have to be seen as an offshoot of an uneven, disparate and bumpy decolonization process. Some of the inter-communal and anti-state violence are a form of insurgency similar to peasant insurgencies studied by James Scott (2009).

This means amongst other things that to correlate state or state-like projection to the formation of security is to ignore the periods of coexistence and peaceful cohabitation among various communities without state interference. In other words, political stability is not just possible through the deployment of centralized governance.

The discourse on the need of an effective state "reoccupation" of northern Mali harks back to colonial and postcolonial strategies of "administration" of outlying regions as opposed to "governance" for urban centers. What this suggestion does is to reinforce a continued model made of a combination of occupation and patronage politics, both of which are caught up in the endemic state, and the irresolution of Malian institutional structures where the army, the bureaucracy and the political elite are concerned, and both of which merely reinforce a sense of exclusion from the "republic" for the populations of northern and central Mali—and increasingly those of the south.

Security in the Sahel and its contenders

Security-based responses to Sahelian problems have not always been the most appropriate or effective, given that these problems have often much to do with poor management of resources, tensions among nomadic herders and farmers, therefore multiple forms of pressure on land and resources. There is an added problem of trust; it is longstanding and structural, and it has to do with the fact that many communities in northern and central Mali barely relate to Bamako and its institutions.[3]

In fact, they have deep suspicions for state officials who often abuse their position of power to extract privileges from vulnerable populations. During the 2012 conflict, some populations in North and Central Mali in fact welcomed a temporary respite from taxation and extortions from state officials. The continued enjoyment of privileges within state structures is hinged upon the sustained instrumentalization of ethnicity by urban elites through clientelist politics. Such privileges include access to lucrative ventures such as smuggling and many forms of extractive practices. In many regions outside of Bamako, the state is seen as disruptor and its representatives as overbearing agents. The state disrupts old hierarchies and only offers patronage in their place while carrying aggressive policies against nomadic populations (ICG 2019). The victimization sentiment is particularly strong among minority populations such as the Fulanis. The sentiment of shared injustice means that Fulanis are increasingly aware of their ethnic identity. In the changing social economy of Sahelian societies, membership into new formations offers an empowering way out of old hierarchies for many socially marginalized individuals.

The Sahel is characterized by forms of livelihood that are articulated around sedentary, semi-nomadic and nomadic livelihoods. Colonial and postcolonial administrations have preferred sedentary livelihoods because they see them as more governable and economically productive. Nomadic populations, however, are often depicted as economically unproductive and environmentally destructive in addition to being antagonistic to peaceful coexistence, social order and centralized authority. These depictions can seriously influence how states and communities manage and mediate access to land, water and other resources as well as negotiate cohabitation amongst different population groups in the Sahel.

It is not surprising then that Central Mali has been part of the larger terrain of conflict. It is part of the Inner Niger Delta which is the most fertile, therefore the most coveted land in the country. In the Macina, a group of Fulani transhumant pastoralists took up arms and refused to pay taxes for access to forage crops and bourgou fields. Their move is a rebellion against both traditional hierarchies, in particular against traditional leaders or dioros, and against state management of access to natural resources (Sangare 2016).

The constitution of self-defense groups in the Macina and elsewhere has to be seen as a response to threats experienced by many Sahelian communities in the face of state repression, the intrusive presence of external actors, religious radicalization and many forms of violence. The resurgence of ethnicity-based militias is framed by both an acute sense of the violence to come and a return to identity as the framework of security. The crisis of confidence in the state and its institutions has been growing. State officials are seen as no more than corrupt predators who prey on vulnerable populations. The context of "the return of identity" as experience, perspective, horizon, refuge and source of claim has thus been the disintegration of both national and "traditional" arrangements.

Tabital Pulaaku and Dogon IZO are two of the most famous ethnicity-based associations. The former was created in the 1990s with Sahelian Fulani from the Diaspora, while the latter, created in 1992, recruits heavily amongst Dogon communities in Central Mali. Their work has been more than just to provide a junction between central government structures in Bamako and rural institutions. Beyond enabling new spaces for the articulation of grievances from an ethnic standpoint, they have also revealed tensions internal to particular ethnic groups. These often have to do with previous hierarchies being questioned, namely old systems of indentured servitude, gender disparities and unequal power structures.

The return to nationalism thus takes up various forms, namely ethnic and sedentary. The year 2015 saw an upcharge in Malian nationalism, which manifested in various expressions of resistance to Tuareg rebels; for instance, the Ganda Koy claim to defend the rights of "black" sedentarized populations against alleged damages inflicted by nomadic communities. Since 2015, retaliations and targeted vendettas have turned into ethnic violence between Tuareg, Songhai, Fulanis, Arabs and other groups that live in north and central Mali (Nga 2017).

Thus religion and jihad now compete with ethnicity as driving force for identity-based mobilization. In fact, militant groups are increasingly manipulating and instrumentalizing ethnic fault-lines. The resulting ethnic pogroms have been far more effective and far more deadly and disruptive than insurgencies in previous years.

The gap between the pursuit of "security policy" and the experience of security by Sahelian populations has meanwhile been widening. Many populations experience insecurity in the form of structural deprivations and chronic precariousness. In addition, the important presence of foreign troops—about 20,000 combined UN/multilateral and French forces—create its host of stress and violence on local populations. New security apparatuses are complex entanglements of geopolitical positioning, economic pursuits, military expansion and predatory moves against weak states. The mechanisms of stabilization and threat containment concurrently deployed in the Sahel through external intervention and state design thus inevitably produce forms of structural insecurity. External interventions encroach upon domestic dynamics.

Instability and violence, in the Sahel in general and in Mali in particular, tend to be viewed exclusively through the prism of the capacity for harm and destruction of armed groups. Rarely, however, do observers examine the parameters of conflict and crisis from the perspective of the insecurity experienced by Sahelian populations.

Gruesome attacks are routinely staged against specific ethnic groups. Ethnic militias have been constituted on the logic of an imperative of retaliation against ethnic "others". As in many other conflicts that go through this kind of attacks, the brunt of the violence is borne by women, children and the elderly.

The enduring insecurity is crippling the region—putting immense pressure on fragile and precarious networks of family and community livelihoods, deepening in the same vein the radical impulses in society's margins and exposing the majority to the vagaries of violence.

International attention has mostly focused on "jihadist insurgency", "violent terrorism", "radicalism", "radical Islam", "rebels" and so forth. The common lens of the terror threat, often applied to the Sahel as a region that displays a concentration of many risk ingredients such as the inability of a weak state to project actual authority on a large territory, endemic poverty, a relatively large population of unemployed youth, the presence of proselytizers that advocate a radical variant of Islam and so on, does not seem to leave room for a plural or nuanced perspective on the Sahelian crisis. However, insecurity takes on forms that aren't easily discernible to external observers unless they take seriously the experience of insecurity by Sahelian populations.

For the majority of Sahelian populations, the issue of security is a dominant concern. An Afrobarometer survey, which involved over a thousand Malians across all regions, shows their most crucial preoccupations are security and sources of livelihood.[4] The survey demonstrates that the crisis caused massive flight in the cities in the North, up to 40% for Gao and Timbuktu in favor of the South; 50% of Kidal's population fled abroad. In this specific city, 88% of surveyed people declared that their businesses had been destroyed and 75% report having been the victim of threats and/or intimidations. Numbers are lower for southern cities, but the general sentiment is one of diminished capacity to sustain livelihoods and an increased condition of exposure to many dangers.

On the whole, the survey shows that Malian populations have a very pessimist view of the future and over the prospect of a realistic end to the crisis. Possible scenarios raised by those surveyed go from fear of Mali's territorial dislocation (38% nationally, 77% for the region of Gao) to the rise in ethnic conflicts in the future (25%). Over half of the people surveyed state having no knowledge about some of the key policy interventions laid out in attempts to end the conflict; 83% state having no knowledge about the Algiers Peace Accords (the main peace agreement, signed in 2015), 54% about the constitutional revision, 49% about the G5 Sahel, 63% about the conference of national entente and 63% about the existence of a Truth, Justice and Reconciliation Commission. These figures show that the degree of disjunction between formal policies and popular experience is staggering. More broadly, they show that "security" is at once a political, economic, material and emotional condition.

But one needs to examine the experience of security and insecurity alongside policies, strategies and mechanisms that have been deployed presumably to restore security and combat threats to security. Their mutually interactive process reinforces a need to rethink security beyond formal pursuits of order and stability. A permanent feature of crisis in postcolonial Mali has to do with various forms of exclusion that go beyond majority politics. Under question here is a problem that predominantly Muslim West African States have been battling with since independence. This problem has to do with a constitutional framework that is heavily informed by French constitutional tradition, one that prioritizes French education as the most legitimate mode of training of postcolonial citizens, a constitutional order that is at odds with preexisting practices of legitimate governance and meaning-making.

Whether the discourse about "marginalization" has been hijacked by clerics and politicians is a question that does not discredit sufficiently enough the claims of groups of people educated in non-francophone systems and systematically excluded from access to state-sanctioned recognition structures. This raises a fundamental dilemma in the current configuration of state-society relations in the Sahel and it is twofold. It has first to do with the breakdown of trust between states and their societies, but also a breakdown of trust among various communities. The hunt for "jihadists" and "militants" has led to the stigmatization of entire communities rightly or wrongly suspected of collusion with imported radicals. Second, the fetishized attachment to elections as mechanism of conflict resolution institutes a state of permanent crisis. Election cycles have become as deadly and as explosive as jihadist attacks. They fragment and weaken social consensus, stoke ethnic hostility and turn former neighbors into enemies. Presidential elections in particular divert resources needed for reconstruction and distract political leaders from focusing on the root causes of crisis. But here lie the contradictions of the liberal peace dilemma in societies in crisis. For the common treatment plan heavily relies on elections as an articulator of transition, the idea that all problems would be solved if only there were "free and fair" elections.

Political demands have become problematized by complex security imperatives that have further entangled the conflict. This entanglement is precisely what is being staged in the current configurations of conflict in central Mali. This takes place in a context, whereby a now long drawn-out war has stymied states that are increasingly beleaguered.

The many wars of the Sahel

To say that Sahelian domestic politics are profoundly internationalized is an understatement. The two combined factors of heavy military response and exteriorization make the Sahel an interesting geopolitical region to study. But the geopolitical perspective requires a degree of historicization as the recent destabilization campaign in the Sahel is a direct outcome of France and NATO's campaign in Libya. The region has been flooded by unknown amounts of weapons coming from Libya and which are serving various groups pursuing divergent interests.

The struggle to keep radical militants out of Europe and the struggle to contain potential migrants within Africa have become entangled. There are two aspects to this question. First, the Sahelian crisis hides many wars. For combatting "terrorists" goes hand in hand with combatting "illegal migration", therefore the containment of African bodies from European soil. It also goes hand in hand with securitizing sources of supply of key African mineral resources to Europe. For this reason, there isn't a single Sahelian conflict but overlapping conflicts serving different purposes.

To safeguard its security, Europe has been pushing its overseas borders into the Sahel. A three-layer border has in fact effectively been erected. The first is a marine/seawall at the heart of the Mediterranean; it includes the redefinition of nautical borders, more rigorous policing of coastlines and the repression of rescue boats at sea. The second is the constitution of North African states into extended, deterritorialized European borders. Sahelian states have become the third layer of this dispositif. The broader intervention for stabilization has systematically abrogated some of the sovereign prerogatives of Sahelian states. Given sustained critiques of the intersection of security interventions and attempts to capture sovereign rights by external actors, the absence of analytical frameworks that explore the combined dynamics of peacebuilding and governance in the Sahel is rather daunting.

In the context mentioned above, many competing logics collide and produce heterogeneous outcomes. There are overlapping security governance agendas, overlapping and competing development and humanitarian mandates attached to these with divergent policy outcomes. The G5 Sahel coexists with the multinational force that brings together Nigeria, Niger, Chad and Cameroon in the fight against Boko Haram. There is also the Accra

Initiative focused on organized crime and "violent extremism"; it brings together Benin, Burkina Faso, Côte d'Ivoire, Ghana and Togo (ICG 2019). At the same time, the Economic Community of West African States (ECOWAS) has been unable to mount an autonomous, coherent, self-funded and effective initiative to strengthen its decision-making capacity and sovereignty. In August 2019, France and Germany announced the establishment of a new partnership—the Partnership for Security and Stability for the Sahel/P3S—to help coordinate European engagement in security and development in the Sahel. This comes in addition to the EU Training Mission in Mali (EUTM) which has been operational since 2013.

In all this, the militarized response has been the most robust, the best endowed and by far the most intrusive policy engagement the region has ever experienced. There are however three problems with the military response. First, the military option colonizes all space for peaceful dialogue. Second, military presence has exacerbated resentment, particularly amongst populations that are feeling victimized with such presence. Third, in the name of restoring peace and order, militarization has sometimes led to the violation of the human rights of the population it is meant to protect.

One of the most important military presences in the Sahel is the UN multidimensional mission. MINUSMA is a mission of facilitation and enforcement of the Algiers Peace Accord, not a combatant mission. It is a weak rather than a strong mandate; it is not a counterinsurgency or anti-terrorist force, yet it is becoming increasingly one. With MINUSMA/G5S, operation Barkhane and national troops combined, there are talks of over 20,000 troops fighting less than 1,000 "terrorists" who do not have planes or helicopters. The scale of the operation is overwhelming, as the intervention logic has come to define just about everything. This in turn has the effect of making the military intervention seem like an irresistible force that subsumes states and regional prerogatives.

While in theory, the successive operations of Serval and Barkhane have firmly isolated mujahideen groups, in reality, a highly fragile context has developed out of an ill-managed crisis. In the North, particularly along the Algerian border, wide swathes of land outside of state control are frequently the site of attacks against state symbols, peacekeepers and civilians. Some of the groups responsible for attacks have begun to expand towards the Centre and the South of Mali. In fact, it is fair to say that a dissonant socio-political landscape has developed out of the post-2012 upheaval; it is a landscape in constant mutation, one which threatens a peace process as well as a national reconciliation framework that is already very fragile.

Further, the WoT has become an all-encompassing paradigm that has effectively set a new order whose logic trumps preexisting security and governance priorities. It has been instituted as a new epistemological order with many consequences for how we can study and understand the Sahel, its societies, its internal dynamics and the forms of intervention needed to resolve its conflicts.

The terrorist paradigm becomes the lens through which everything is filtered through, even if not all violence is of a terrorist nature. A common understanding of violence in the Sahel posits states and their allies as producers of security and of non-state actors as producers of insecurity (Niang 2018). Such simplistic binary reading however deflects from the thorny work of tackling the deep-rooted problems of the postcolonial African state and increasingly complex political and economic motivations of transnational engagements.

The security imperative, especially as conceived from the perspective of the WoT, has become the main source of security ordering. It orders state policy, the relations of the state with its (security) partners, especially multilateral agencies and donor countries. It also conditions the articulations of state-society relations.

Both the WoT perspective and the multilateral force are framed as order-producing interventions. Interventions are geopolitically driven and ideologically informed. They push forward a vision seen as normatively progressive and capable of transforming modes of social interaction. In the last few years, there has been an explicit alignment for instance of technology with policy shifts; the goal for interveners is essentially to minimize loss of soldiers and to distance from targets of strikes. France's engagement is a case in point. There is a contradiction in France's desire to disengage from the Sahel while maintaining control on the region. If anything, its military presence has recently been increased with the approval of the heads of states of the G5 Sahel who had been summoned in Pau, France in 2019 to rubberstamp a well-orchestrated policy decision. However, securitization politics is doomed in advance. The geopolitical interests are simply colossal.

In December 2019, France announced that it had successfully used drones in areal strikes in the Sahel from its Nigerien base. Drone warfare has become a crucial brand of security pragmatism, especially in places where there is little to no democratic control over the use of technology in combat. China for instance is the third provider of small weapons in Africa and it has troops stationed in the Sahel as part of MINUSMA. What African security will look like in the 21st century and the years to come has very much to do with European and American anxieties around controlling strategic resources—a matter of life and death, as intervention elsewhere in Iraq and Libya has shown. It is also a war for positioning in the coming reordering of the world. Competing actor engagement means that there is little room for African initiative. The fact however that there is an absence of long-term, coherent, consistent approach to Africa should not always be seen as a bad thing as it gives space for initiative to African states.

The multifaceted intervention amounts to a security governance framework that encompasses policies, technological experimentations and institutional reforms. Governance and critical security studies analytics have however denounced the nefarious consequences of the technological argument in particular where the security governance logic has led to the creation of "cities under siege" (Graham 2011).

This brings me to the notion of CNWs as exemplified in the Sahel. My argument is built upon Kaldor's (2012) conceptualization of new wars in the post-Cold War era. Like Iraq, Syria and Libya, the Sahelian conflict is a CNW in its causes, its manifestations and its effects. It is a conflict that emerged at the confluence of a fraught, incomplete decolonization process, geopolitical struggles for control over strategic regions and their resources by powerful states, and post-Cold War reconfigurations of political regimes and global re-alignments. For this reason, the conflict that started in Mali cannot be studied as just another ethnic conflict or civil war. In fact, in the past few years, external factors, from jihadist militancy to external multilateral interventions, have been most significant in imparting the course, the scope and scale of the conflict.

What to make of all this for researching complexity of conflict

The notion of security and the liberal peace paradigm

The successful implantation of liberal institutions and procedures of resolution (elections, DDR, transitional justice frameworks) have created for their proponents an illusion of distinct zones of peace and zones of war, when in reality, these are merely the "effects of mutually constitutive international political, social and economic relations" (Barkawi and Laffey 2001: 2). In essence, the pursuit of the liberal peace dogma has jeopardized the principles of reciprocal restraint that could otherwise have sustained, in places, the gradual transformation of the means of pursuit of political claims from violent to peaceful means.

The security policy reform activated in the DDR process and counterinsurgency strategies is based on the idea that anti-state violence is informed by primitive, irrational and pre-modern thinking and needs to be fought using technology-based, sophisticated and legal instruments. Both local and international actors share this thinking. Thus, the construction and reproduction of insecurity imaginaries justify securitization while serving to neutralize criticism of analyses of "experience" as the stuff of research.

I'm not talking about "vernacular security" as a category of account or self-narrative, but the very experience of insecurity as generative of a particular kind of cultural, if not ontological anxiety. A question that arises out of this is the question of theorizing security in places of conflict beyond commonplace understandings of security as an "achieved" place of "peace". There is a gap between paradigms commonly used by scholars and policymakers and the actual internal security concerns in the Sahel.

Geopolitics as structure

The current configurations of the conflict are a result of both immediate factors, namely those that led to the 2012 impasse, but also more profound factors to do with longstanding simmering grievances, increased competition over dwindling resources, partisan political and factional conflicts, conflicts of status and power with structured groups and so on. The conflict itself has mutated over the years, and this is reflected in the formation but also the changing demands of armed groups involved. Whereas on the one hand the Tuareg have been demanding statehood explicitly since 2012, the Fulani group of MLF has been seeking to set up a Caliphate, while on the other hand the Dogon have been reaffirming their rights to ancestral land. For all these reasons, the conflict in the Sahel cannot continue to be studied as one of a terrorist assault on fragile states.

Further, the three most recent iterations of the Malian conflict are all directly linked to regional and international phenomena that have to be seen as more than just amplifying factors. First, the 1990s conflict was an extension of the Algerian civil war. In 1991, the French supported the coup that toppled the democratically elected Front Islamique du Salut in Algeria. If subsequent campaigns of repression and the massacre of hundreds of thousands of Algerian civilians was not the responsibility of the French, the implications of their support to the repressive Algerian state cannot be ignored. One needs to remember that the North of Mali is also the South of Algeria. Understandably, Algeria seeks to maintain its security forces in this region.

Second, the wave of Islamist radicalization from 2003 onwards was the outcome of the expansion of the frontiers of WoT from Central Asia to the Sahel. Until after the aftermath of 2001, the Sahara-Sahel region was considered marginal as a global "security threat". This changed emphatically around 2003. Third, the 2012 reigniting was a spillover of the destruction of Libya and the murder of Mohamed Qaddafi. The disintegration of clientelist networks, the kind over which Qaddafi and Compaore held sway, were central to the postcolonial status quo in West-Central Africa. Like Houphouet Boigny of Cote d'Ivoire before him, Compaore was happy to play the middleman who performed contradictory roles to the successive conflicts in the region as it suited his own regional pursuits. The crumbling of these networks coincided with the resurgence of dissident movements in Northern Mali under new leadership.

Since the early 1990s, the "Malian conflict" no longer was a conflict strictly defined by dynamics internal to Mali or confined to Mali—and it never really was—but rather a multifaceted, sprawling crisis entangled in broader regional and global processes that also frame "local dynamics". This means at the very least that a study, for instance, of the Tuareg question, must necessarily take a perspective that is at once historical, comparative, regional and transnational.

The larger point here is that the Sahelian conflict should encourage a rethink of the changing nature of the geopolitical struggle. Geopolitical wars are being outsourced; they should compel us to rethink the kind of knowledge that is enabled or deactivated by related emerging orders. The geopolitical dimension of new wars cannot be overstated. At the same time, an enduring history of western interventionism in the Sahel makes the issue of responsibility a complex one.

Africa is a place of experimentations

The notion of a "new scramble for Africa" conceals a larger multi-sided enterprise of experimentation, from the deployment of military technologies to expanded corporate enclaves now increasingly folded into militarized practices. The French Special Forces are, for instance, responsible for the protection of the French uranium giant AREVA.

In a conflict that has become a perfect storm that is picking up strength rather than declining, it becomes difficult for a researcher to capture the complexity of a social phenomenon. Implied in the discussion above is the importance of unpacking entangled positions historically and in war contexts like the one in which the Sahel is engulfed. The researcher himself/herself is not an observer removed from this entanglement. The positionality of the researcher implies a degree of hierarchy involved in defining and imposing concepts and a framework on any given situations. Armed with sophisticated methodologies and measurements tools, a researcher should try as much as possible to put these to the test of context rather than try to fit a conflict context in their preset research design. CNWs like the Sahel require an involved approach to field research despite the conflict conditions, given the sheer complexity and the rapidly changing conditions and shifting alignment of actors involved.

It is also important that we circumscribe the meaning we assign to concepts and ideas, and also become more cognizant of their different uses and manipulations by scholars and policy-makers alike. In the specific context of the Sahel, external intervention is based on overlapping security paradigms that one needs to disentangle: from Lyautey and Gallieni's « peaceful penetration », David Galula's counterinsurgency policy, to American antiterrorism and use of "low intensity conflict". Bagayoko (2020) contends that African policy-makers themselves "use, make reference to, or align themselves on norms and analytical categories which are insufficiently contextualized". Long before Bagayoko, Stephen Graham (2011) had called attention to the "military-corporate-think tank complex" that feeds the WoT in all aspects, from working ideas and key concepts to advanced drone technology.

Policies can expand or conflate the remit, or reinvent and repurpose concepts. Thus, "intervention" has inevitably become an all-embracing and boundlessly flexible concept. In the same manner, technology—from the use of surveillance equipment to the use of drones as weapons of destruction—hides a vast range of new possibilities. The technological dimension thus obscures an important emergent trend in both policy and practice that has the capacity to stretch the meaning of concepts.

Humanitarianism is another concept whose use has become entangled in new geostrategic alignments. Where in particular policy-making is framed within the strategic logic of warfare, David Kennedy (2004) shows that humanitarianism has become an all-encompassing concept and a mantle that shrouds destructive practices. In other words, the vocabulary of necessity, proportionality and humanity does not translate a face value understanding of these terms' moral and ethical source. Like the notion of "security" and cognate concepts, the notion of "humanitarianism" needs to be constantly assessed for accuracy.

Conclusion

A great confusion persists in relation to analyses of the Sahelian conflict. There is a difficulty to distinguish between ethnic grievances, sheer banditry, Islamist radicalism, competition over resources and local vendettas that are constantly reignited in times of conflict. The conflict itself has mutated over the years and this is reflected in the changing demands of groups involved: the Tuareg demanding statehood and the Fulani seeking to set up a Caliphate and so on. However, the WoT logic continues to determine the dominant narrative on the conflict.

The geopolitical dimension of new complex wars such as the one in the Sahel cannot be overstated. In Mali and elsewhere, the geopolitical imperative defines conflict resolution strategies, the mode of intervention, the preference for militarized rather than political solutions and so on. The G5 Sahel in particular is a reflection of contradictions laid bare by geopolitical wrangling. It reflects coloniality and ideological continuity of a clientelist logic that has maintained strongmen for decades in West Africa, often associated with the infamous Françafrique network. West African states, North and South, share the same vulnerability and tumultuous internal politics.

In this context, the ethnicity angle is increasingly proving to be both an incomplete and misleading framework. No single-factor analysis is in fact able to account for the complexity and the history of conflict. For this reason precisely, the militarized response to the crisis seems both illogical and counterproductive. "Ethnic conflict" as such is but a mythologization in presenting a part-as-totality. The first thing is therefore to rethink the study of "ethnic conflict" beyond the question of ethnicity alone. There are three challenges clinked to this effort of deconstruction. The first is to take stock of accumulated lacunae on representations of wars in "far-flung" places. These carry an ill-conceived understanding of social and cultural boundaries in West Africa as well as blurred distinctions between "zones of war" and "zones of peace".

The second thing to bear in mind is the fact that there is a collision of worldviews when it comes to how security is conceived by various actors and protagonists in the Sahel. The WoT has however turned the Sahel into a new frontier of man-made disasters. There is clearly a need for a reconceptualization of security on the basis of the experience of recipients of interventions, their historical conditions and the purpose of their struggle. More specifically for the Sahel, one needs to pay careful attention to generalized circumstances that deny social value to Sahelian lives and the deteriorating structures of living that no longer allow predictable social patterns.

Notes

* This chapter was written before the August 18th, 2020 military Coup in Mali.

1 International Crisis Group, Mali: la paix à marche forcée" 22 mai 2015, https://d2071andvip0wj.cloudfront.net/226-mali-an-imposed-peace-french.pdf

2 Fulani dignitaries went to Bamako to request arms from the defense ministry but were turned away

3 Mutual accusations are rife: for herders, agricultural land has been expanding to the detriment of nomadic herders; for farmers, herders have not been respectful of established codes and do not limit themselves to designated *burti* or trails, but rather let their cattle trample farming land.

4 "Résumé des Résultats. 7ème Tour de l'Enquête Afrobaromètre au Mali, 2017" Groupe de Recherche en Economie Appliquée et Théorique (GREAT).

References

Bagayoko, Niagalé. (2020) "Sahel: Il faut repenser la sécurité du continent selon une perspective stratégique africaine" https://www.lemonde.fr/afrique/article/2020/01/10/sahel-il-faut-repenser-la-securite-du-continent-selon-une-perspective-strategique-africaine_6025461_3212.html

Barkawi, Tarak and Laffey Mark (eds.). (2001) *Democracy, Liberalism, and War: Rethinking the Democratic Peace Debate.* London: Lynne Rienner Publishers.

Graham, Stephen. (2011) *Cities under Siege: The New Military Urbanism.* New York: Verso Books.

ICG. (2019) "The Risk of Jihadist Contagion in West Africa" Briefing 149/Africa, December 2019.

Kaldor, Mary. (2012). *New and Old Wars. Organized Violence in a Global Era.* Cambridge: Polity Press.

Kennedy, David. (2004) *The Dark Sides of Virtue: Reassessing International Humanitarianism.* Princeton: Princeton University Press.

Nga, Armelle. (2017) "Mali: risques d'un conflit inter-éthnique dans le centre du pays" *Africa News*, 24 March 2017 http://fr.africanews.com/2017/03/24/mali-risques-d-un-conflit-inter-ethnique-dans-le-centre-du-pays/

Niang, Amy. (2018) "Stateness and Borderness in Mediation: Productions and Contestations of Space in the Sahel" *Adaption und Kreativität in Afrika, the German Research Foundation Working Paper N26.*

Raineri, Luca. (2019) "Fous de Dieu ou indignés? Crise de gouvernance, polarisationsidentitaires et populisme religieux au Mali" In Baz Lecocq and Amy Niang (eds.), *Identités Sahéliennes En Temps De Crise: Histoires, Enjeux Et Perspectives.* Berlin: Lit Verlag, pp. 203–226.

Sangare, Boukary. (2016) "Is the Mali Centre an epicenter of djihadism?" *Note d'Analyse, GRIP*, 20 May 2016.

Scott, James. C. (2009) *The Art of Not Being Governed: An Anarchist History of Upland Southeast Asia.* New Haven: Yale University Press.

Sogge, David. (2011) "Weak States and the Savage Wars of Peace" *Global Dialogue*, 13(1), Winter/Spring2011.

The Biarritz Declaration for a G7 and Africa Partnership, 26 August 2019. https://www.elysee.fr/en/g7/2019/08/26/g7-leaders-declaration.

8 Researching women, unpaid care work, and peacebuilding in Darfur

Fatma Osman Ibnouf

Introduction

Care arrangements during wartime are complex, and include a range of tasks that are essential for the survival and the well-being of individuals, families, and communities affected by war. Unpaid care work is an important aspect of these activities. Unpaid care work arrangements include all the functions performed by women during wartime and in post-conflict contexts. These arrangements are the foundation which save lives and sustain communities, as can be seen in the case of the conflict in Darfur. The Darfur conflict, which started in 2003, was a great source of suffering for women and others. The war increased the demand for care in quantitative terms (prevalence of disability and illness, number of injured people, orphaned children, etc.) and in qualitative terms (dealing with extreme emotional situations such as sexual violence, trauma, and grief). In addition to general suffering associated with war conditions, many women were raped or witnessed the rape of their beloved ones, including daughters, sisters, and mothers. As the main providers of care to others, women in Darfur have borne the brunt of stress, trauma, and drudgery associated with sustaining individuals, families, and communities during and after war.

Researching and understanding the lives of women care providers in a conflict-affected region, like Darfur, come with significant contextual, conceptual, and methodological challenges. Apart from finding relatively safe and conducive field research sites that would yield useful data, the researcher needs to consider the conceptual framework and methodological approaches that would best convey the experiences and voices of women. Adopting a feminist perspective, especially being empathic of the complexities of the lives of the women, is valuable in navigating some of these challenges. A feminist perspective entails the consideration of women's subjectivities, both in times of war and peace. It considers the different ways in which women conceive of themselves, individually and in relationship to others, and their roles in society at different times, especially as caregivers.

This chapter reconceptualizes care work as a specialized activity in wartime that is difficult to research, but one that makes women a specific kind of protagonist whose knowledge and perspective is necessary to understanding the structures that contribute to the survival and well-being of the communities before, during, and after conflicts. It posits that a feminist perspective has the added benefit of challenging misunderstandings of the various roles performed by female caregivers in building and sustaining peace. It translates theoretical frameworks into concrete analysis, and puts on center stage the knowledge and experiences of marginalized women, in this instance, Sudanese women in displaced camps. A feminist perspective also underlines the need to infuse the practical experiences of the affected population in a way that often mirrors real life to bring about more reflective, effective, and innovative peacebuilding processes.

The reflections and data in this chapter are drawn from a larger study of women's care work carried out from August to September 2016 in five internally displaced people (IDP) camps in the states of North Darfur and South Darfur in western Sudan. Mixed quantitative and qualitative methods were used to collect structured and unstructured data.

Darfur: researching caregiving in a complex conflict context

The Darfur Region poses particular challenges for a researcher. It occupies the far west of Sudan and shares international borders with four countries: Libya, Chad, The Central African Republic, and South Sudan. Darfur is one of the larger and more densely populated regions in Sudan and covers an area of 493,180 square kilometers. It is divided into five federal states, namely Central Darfur, East Darfur, North Darfur, South Darfur, and West Darfur. The region's population is estimated at 9,241,369 in 2017 (Central Bureau of Statistics, 2017). The Fur, after whom the region is named, is the largest ethnic group in the region (Ali, 2014). The two largest cities in the region are El-Fashier in North Darfur and Nyala in South Darfur. The population consists of a diverse mix of ethnic groups: the Fur, Massalit, Daju, Borgo, and Tama, which are the largest agriculturist groups in the region; the Rhezeghat, Zaghawa, Habania, Beni Hussein, and Zeiyadiya, which are mainly sedentary cattle herders; and the Beni Helba, Beni Jarrar, and Mahameed, which are mainly nomadic and semi-nomadic groups. Despite the presence of a plethora of indigenous languages, all of the groups generally speak Sudanese Arabic. They also have a shared religion (Islam) and a share of the traditional culture, which is patriarchal in nature.

Much has been written about the origins, the nature, and the impact of Darfur conflict. The Darfur conflict has deep historical and environmental roots, but it has changed over time and increasingly became "militarized" by the first decade of the 21st century. Since the great Sahelian drought of 1970, more pressure has been placed on the savannah drylands of the region, pushing cultivators from drier near-desert north Darfur to resettle in the rich savannah of south Darfur (Abdul-Jalil, 2008). Desertification, population movements, and the accumulated effects of human activity generated local conflicts around access to and rights of use over natural resources, mainly land and water (ibid).

Starting in 2003, Darfur experienced cycles of armed conflict between the Sudanese national government, the Justice and Equality Movement (JEM) and the Sudan Liberation Movement (SLM/A). JEM and SLM/A rebels are both drawn from members of the Fur, Masalit, and Zaghawa ethnic groups. The two main rebel groups (JEM and SLM/A) gradually fragmented into several factions, and nowadays, more than 40 rebel factions exist in Darfur. There have also been intra-rebel clashes. There are also many small unidentified armed groups that attack villages, kill civilians, rape women, rob and loot livestock, and hijack humanitarian vehicles. In addition, Darfur is a theatre of ongoing and unresolved conflicts between sedentary farming and nomadic-pastoral communities. This has produced a complex network of inter-related conflicts.

International media reports as well as some scholarly work have tended to portray the Darfur conflicts as primarily a struggle between African farmers and Arab herdsmen over land and water. Yet, there are complex root causes and proximate causes, encompassing contentious historical legacies, oppressive tribal and social structures, and clashing local norms as well as competing economic and political claims. Conflicts in neighboring countries, namely Chad, the Central African Republic, and Libya, have facilitated violence in Darfur. Hundreds of lives have been lost in the violence. The different protagonists in Darfur signed a series of ceasefire agreements on humanitarian access and security, but have frequently broken them. In fact, the many Darfur Peace Agreements (DPAs), with their different shortcomings, have heightened rather than lessened the conflict at different times.

The deployment of United Nations-African Union Hybrid Operation in Darfur (UNAMID) in 2007 has helped to improve the security situation in the region. However, the humanitarian situation continues to be grim for millions of Darfurians. More than two million people are still living in IDP camps, according to recent estimates from the government of Sudan. Displacement has adversely affected all aspects of the lives of these people. Although both men and women face considerable challenges when displaced, women experience a great deal more. Roughly three-quarters of the IDPs are women and children. Many women have been raped, abused, and experienced gender-based war crimes. Child marriages, domestic violence, maternal deaths, and other incidences of social exploitation have been part of the experiences of females in the IDP camps. Even when these women do not have access to support services or care, they are still expected to be the primary caregivers in the IDP camps because Darfurian patriarchal culture places the responsibility and burden of care on women.

Even beyond Darfur, care is a strongly feminized field of work. As Liera and Saraceno (2002) point out, care work was invisible when it was interpreted as something intrinsically female. Feminization of care means that women are automatically expected to perform the role of caregivers. In many "traditional" cultures, women as mothers, sisters, wives, daughters, and in-laws are supposed to take on far more care work and domestic activities than men. Care, well-being, domesticity, and practices of familial responsibilities are inextricably linked in the socialization of women in these cultures.

Care work, which is mostly uncompensated in these contexts, has important implications for the well-being of individuals and households (Budlender, 2010). During the Darfurian conflict, the disproportionate burden of unpaid care work has largely fallen back on women and girls, many of whom are also in need of care. They have to deal with their own war-related trauma, sleep deprivation, and limited resources, while responding to stressful experiences, including birth, assault, rape or the unexpected death of loved ones. Despite the considerable difficulties when dealing with the suffering of the injured or the dying, the women caregivers in this study experienced great satisfaction when relieving pain or treating injured people or assisting with childbirth.

The experiences of the women recounted in this chapter are indicative, not comprehensive, since it was not possible for the researcher to visit all IDP camps in Darfur. The number of IDP camps in Darfur exceeded 20 at the time of the field survey. Security restrictions made it dangerous to access many IDPs camps. Zamzam and Abu-Shook in North Darfur State and Otash, Kalma, and El-Sereif in South Darfur State represent the five largest refugee camps in Darfur. Zamzam is the largest IDP camp in the whole Darfur region. The Zamzam IDP camp is an open camp that currently hosts over 200,000 people. Many of the IDP camps are slowly becoming permanent settlements. The government of South Darfur State wants El-Sereif camp to become "El-Sereif City" as part of a larger plan to transform some camps into permanent towns. The IDP camps are administered by a combination of traditional, government, and international authorities. The Sheikhs are the IDP camp leaders and community representatives to government officials, United Nations agencies, including UNAMID, and non-governmental organizations. Women are part of the camps' consultation committees, although their roles are rarely equal to that of their male counterparts.

Darfurian womanhood, care work, and feminism

Darfurian womanhood and care work

Almost all Darfurian ethnic groups are structured on the basis of patriarchal culture. Womanhood has a devalued status in Darfur, and this is evident from the expectations and attitudes from within families and in the larger society. Women are well aware of their care and domestic work responsibilities. Men and boys are not expected to shoulder the same care work responsibilities. Girls usually internalize these expectations from their mothers, and they are socialized to fit into the "ideal womanhood" model and to be able to "take care of the family affairs". They are told to respect their husbands and endure their marriage as women who are meant to care for their family even in wartime. Womanhood is thus conceptualized on how women act according to the customs rules and expectations attached to "women's work" as caregivers. Carol Gilligan argues *In a Different Voice* (1982) that women judge themselves according to their ability to care.

The response of Darfurian women to the realities of their situation in IDP camps is consistent with their ability to cope throughout life in their socially ascribed gender roles and responsibilities. Gendered division of labor in Darfur and the Sudan means that women are responsible for the more time-consuming and household intensive tasks. Men hardly perform any domestic activities, during or after war. Domestic activities are generally not compensated, classified as being productive work, or included the national statistical accounts. Women's caregivers are therefore not considered part of the wage economy.

As caregiving demands escalate during and after wartimes, so do the expectations and responsibilities of women. Women do all that is necessary to maintain society, from caring for children, elders, relatives, the injured, the disabled, the dying, and themselves, to finding sustenance for their family members. Women perform caregiving with neither the resources nor the institutional support required to help them through adverse conditions. Typically, in post-conflict settings, in addition to carrying on with their traditional roles, they take on breadwinner roles because the men are either in the battlefield, dead, captured by the enemy, or trapped in their homes without access to work. Displaced women continue performing their care work while they are still dealing with the war-related trauma and experiences of loss/grieving.

Defining care and care work

A careful reading of the literature suggests that care work varies in meaning, according to the context of use in the different disciplines. According to Peterson, it refers to all the necessary activities to reproduce human life at personal and community levels where human life is generated and daily life is maintained (2002). Tronto (1993) sees it as a general activity, which includes collective care rather than a dyadic one existing between two people – a caregiver and a dependent care-receiver – which Carol Gilligan (1982) and Noddings (2002) envisaged as being emblematic of the caring relationship. Tronto (2013) observes that caring very rarely happens only between a single caregiver and a receiver, but is more complex, even in the space of the family, where multiple people are caring for each other. In *The Heart of Justice: Care Ethics and Political Theory,* Daniel Engster (2007) defines caring as "everything we do directly to help individuals to meet their vital biological needs, develop or maintain their basic capabilities, and avoid or alleviate unnecessary or unwanted pain and suffering, so that they can survive, develop, and function in society" (cited in Friedman, 2009). Some scholars emphasize the moral dimension of caregiving, noting that it is a practice of empathic imagination, responsibility, and solidarity with those in great need. For Kleinman, it is a moral practice that makes caregivers, and at times even the care-receivers, more present and thereby fully human (2009). It connects people and imbues these connections with a wide range of emotions (Lawson, 2007).

In this study, I adopt the concept of "care" put forward by Orozco and Leiras (2017), namely activities that serve, on a daily basis, to enhance or sustain life; the life of all people as interdependent subjects and in any social context. In Darfur, it includes all the functions performed by women as caregivers in conflict and post-conflict situations. It involves their quotidian arrangements for the provision of basic necessities to ensure the daily survival, emotional and psychological support, care for children and adolescents, and protection for family members. As Bakker contends, care work includes all the essential tasks, for example, childcare, cooking, and cleaning, required to support the development of cognitive, emotional, and physical capabilities of people (2007). In short, care is embedded in everything and everything worth something takes preparation. Care activities are essential for the child growth, family survival and well-being, and community solidarity. Care work involves offering physical, psychological, and emotional support, in addition to protection and family relations maintenance. It has an associated emotional response and it influences how people act towards others around them. In all its forms, care work benefits the society on the whole (ILO; 2007) by providing the daily material, emotional, and social support upon which survival, work, and social life depend.

A feminist ethics of care

Embedding the study of unpaid care work within a feminist perspective centers the roles of Darfurian women in sustaining and reproducing life and society in a major conflict setting. It also foregrounds how the work of these women reflects an ethic of care defined by obligations, responsibilities, and relationships arising from personal commitment, social expectations, and expediencies of wartime. Sevenhuijsen et al. suggest that understanding this ethic of care should start from the ground knowledge of how responsibilities for caring are actually practiced and from deliberations about how these responsibilities can be supported and enhanced (2003).

A feminist ethics of care addresses the feminization and devaluation of caring practices of women in Darfur and other conflict zones. Worldwide, women typically spend disproportionately more time on unpaid care work than men (Ferrant et al., 2014). Feminist economists have led the call for unpaid care work to be "counted" in statistics, "accounted for" in representations of the economy, and "taken into account" in policymaking (Budlender, 2010). It is impossible to talk about care without considering the emotions, moral duties, the processes of social interaction and subjective formation, and the intimacy and corporeality of these relationships (Orozco and Leiras, 2017).

In the Darfur case study, a feminist ethics of care provides a suitable theoretical approach for research and post-conflict peacebuilding for four reasons. First, women bear the burden of unpaid care work, responding to the cultural expectations of their social positions and responsibilities. A feminist ethics of care approach brings to light the challenges women caregivers face and contributes to broaden thinking on the linkage between regular care work and wartime care work. It opens up possibilities to address the subordination of women, the disproportionately higher level of unpaid care work they provide, and how care work has impacted their life. From this we can imagine a case to be made for greater attention to the need to taking care work into account, to advocate "namely recognition, reduction, redistribution and representation". Second, a feminist approach provides an effective means to deal with the common devaluation of caring arrangements. It insists on the consideration of women's role in unpaid care work arrangement to make the invisible visible. Third, the approach recognizes that voices of women caregivers have not been heard in the conception and implementation of peacebuilding processes. Therefore, the critical role that women play through their care work, it should be argued, qualifies them as leading voices in peacebuilding. Fourth, a feminist perspective offers useful insights in assessing whether peacebuilding frameworks sufficiently recognize the practical realities of women's lives, especially the challenges of providing care in and out of conflict. It would enable such frameworks to adequately address gender equity and gender justice, and to advance women's human rights concerns and access to justice.

Key aspects of the feminist care theory are used as "pillars" for the Darfur case study. A feminist ethics of care perspective allows for the consideration of the experiences of those affected by war in a meaningful way, a matter of concern for women caregiver in Darfur. During and after war, women expand their role to arrange for all types of life-saving care work. Thus, the care burden is multiplied exponentially under such situations.

Data collection and conducting research in a conflict situation

Many hardships, difficulties, and challenges are involved in conducting research on the lives and work of women in war-affected areas like Darfur. As a researcher, I was aware of the security, ethical and methodological complexities and restrictions associated with conducting fieldwork with people affected by war and displacement. It required patience and careful preparation for unexpected problems. Before I commenced my field research in September 2016, I had narrowed down my scope of research to five IDP camps: Zamzam and Abu-Shook in North Darfur and Otash, El-Sereif, and Kalma in South Darfur. Apart from the regional differences, each of these camps has its particularities. Despite all these preparations, I still expect the unexpected during my fieldwork, including the threats to my personal safety.

To capture the complexity, diversity, and plurality of the women's experiences, I utilized a judicious mix of qualitative and quantitative data collection methods in the field. I used field survey, structured interview (in pilot phase), semi-structured in-depth interviews, informal interviews, focus group discussions, and storytelling. My research assistants, Nafessa, Hawa, and Fatima, spoke the local languages and were formerly displaced persons. They were familiar with the area and the people, and accompanied me throughout the field research process to assist with comprehension. Before the beginning of the research, they were properly briefed on how to facilitate the research process, especially focus group discussions. Interviews and notes were done in Arabic and were then translated as precisely as possible into English. I was equipped with a digital recorder, field notes diary, and video recorder. I took photographs and tape-recorded videos for accuracy, with the participants' consent.

Even with the best planning, entering IDP camps in Darfur was not always easy. For example, I managed to enter IDP Kalma camp, despite the high levels of insecurity in the camp. Even some government officials are unable to enter the camp, and refer to it as a political platform for the SLM. Despite all these challenges, my experience reinforced my fundamental determination to bring to light the fact that women's unpaid care work during and after war tends to be perceived merely as an extension of their household duties.

While having safe and secure access to the IDP camps was significant, perhaps the most significant step in my field research was gaining the trust and confidence of the participants. This meant spending time in the displaced camp, observing and interacting with potential participants, and becoming slowly integrated into the setting. This enabled me to create meaningful rapport with the women in the camp and to maximize to their involvement in the research, while protecting their right to refuse. I ensured that no one came to an interview blindly or would feel pressured to speak up.

I made sure that the women identified as participants understood the nature of the research, and willingly consented to participate in it. They were asked to indicate their consent verbally (for the illiterate) or by signature after reading the consent form (for the literate women). They were also given the option of consenting to or declining audio and video recording of interviews; some declined and the data collected was recorded by handwritten notes. The transparent and sensitive manner in which the participants and I worked out permissible boundaries of the research opened up opportunities and expedited the establishment of a good working relationship. It also created a conducive environment to discuss all issues related to caregiving during and after war.

The lack of available, accurate statistics made it virtually impossible to compile a clear sampling frame beforehand. When I tried at first to sample randomly at the camps, I encountered a number of difficulties. Some of the women refused to take part in the research activities. I had to visit the IDPs local leaders and urge them to identify people who would be willing to share their knowledge. With help from local leaders and the research assistants, I described the purpose of the study to the women. As a result, more women in the IDP camps expressed willingness to participate in the study.

Interviews were carried out inside IDP camps to ensure that interviewees felt comfortable and I maintained strict confidentiality. Any interview process with war-affected people has its limitations, and getting good data depends on the researcher's experience and wisdom. Recounting what happened during wartime was very painful for the Darfurian women. My best interviews were done in a conversational rather than a mechanical question and response method. This technique, which frames the process as an exchange rather than extraction, built rapport and put the participants in a relatively comfortable situation. I also raised difficult questions towards the end of the conversation, by which time the participants had adequate comfort to address them.

During interviews, I gave each participant enough time to answer the questions, as they tried to recall their day-to-day painful experiences and details of what had happened during the war. I listened not only with my ears, but also with my heart. Listening to the crimes against some of the women, which were horrendous and humiliating, was draining and overwhelming. Some of their eyes were filled with tears when they described the deaths of family members, relatives, neighbors, and friends at the hands of the rebels or military forces. I found myself sometimes emotionally involved. Empathy is especially important when dealing with such cases. I expressed genuine concern for their issues and treated them carefully. Through body language, eye contact, facial expressions, attentiveness, I tried to express my empathy and care.

The interviews were triangulated with focus group discussions. I conducted two focus group discussions in each IDP with help from my research assistants using the same question guide. The group discussions in the five IDP camps in Darfur focused on whether women considered that they have relevant knowledge and ideas on what should be done during the peacebuilding process, including activities, needs, roles, and leadership. The women also considered and discussed whether they felt that their pivotal role in care arrangements is unjust and the responsibilities of men as well as NGOs, INGOs, and CSOs.

Conducting fieldwork on "wartime caregiving" in "Darfur" is one of the best and transformative experiences that I have ever had in my academic journey. I learnt a lot about the different approaches that women used during wartime to save lives, relieve suffering, and care for their dependents. I observed various physical and emotional processes, including tasks like dressing, feeding, toileting, managing medication, and mobility. I witnessed how women, in the absence of state healthcare services, used natural remedies like fennel (*Foeniculum vulgare*) seed to relieve constipation in newborn children. Women recounted how during wartime, they applied crushed garlic cloves or lemon powder on wounds to stop bleeding, help the healing of wounds, and even treat skin affections.

Being in the field illuminated the difficult conditions and limited resources with which women performed everyday care. There is no electricity in IDP camps. Indigenous foods have continued to be the "foods of survival" in the camps. Women used simple and low-cost traditional food processing techniques (fermentation, salting, smoking, drying, etc.) to increase the shelf life of the food products. They collected uncultivated wild food products such as kawal *(Cassia obtusiflora)* and green leafy vegetable (*Sonchus* spp.), and processed some inedible wild products into conventional food products like makhait (*Boscia senegalensis*). Edible insects such as desert locust (*Schistocerca gregaria*) and birds such as red-billed quelea (*Quelea* quelea) were also used as sources of emergency foods during and after wartime.

I also learned some painful truths about the violation of women's rights in IDP camps. Conditions in the IDP camps are bleak and women are exposed to great risks of violence. There is enough pain to create a much louder outcry than exists in the media or by the international agencies. With low-resource settings, a limited income capacity, and lack of basic services – in the absence of relief aid, violation of their human rights, domination of cultural practices, and patriarchal systems and challenges of displacement continue to escalate – the women still worked round-the-clock as caregivers for their dependents. Despite all these challenges, in the words of "Khadija", a 32-year-old widow and mother of five, "caregiving is a labor of love".

The data collection and fieldwork, true to feminist methodologies, was not a one-way street. It was also beneficial for some of the women who participated in the study. Almost 90% of them indicated that it was a positive experience. One of the respondents stated after an interview: "I feel much better having talked with you … I would therefore like to thank you very much for listening to my experiences". Another added: "I have a desire to make things better for my family and community after I talked to you … I'd really like to contribute my part in building peace". For my research assistants, Nafessa and Fatima, it was also an empowering experience. It made them more conscious of their social role as caregivers and committed to advancing the interests of displaced women. They both represented displaced women in the Darfur Internal Dialogue and Consultation Conference held in Khartoum on 30 October 2017, in a departure from normal practice. In previous conferences, representatives were selected from the State and Localities officials to speak on behalf of IDP women.

The experiences of Darfurian women through different data collection methods

Individual interviews: when state and non-state actors fail

Feminist methodologists have paid particular attention to different types of data gathering techniques and their possibilities, a number of feminist scholars have elected to use in-depth interviews as preferred method (Harris et al.,

1995; Oakley, 2000 cited in Pini, 2002). The key informants included omudas and sheikhs, who were purposely selected for their knowledge of residents, ethnic diversity, and their services in the IDP camps. They also had helped identify participant experts for the study.

One of the main threads in the individual interviews was to understand how the women acquired basic necessities and resources for themselves and their families. During interviews, some of the key informants indicated that some humanitarian aid organizations such as World Food Program (WFP) provided monthly provisions of basic necessities to support life of IDPs. However, during fieldwork, the surveyed samples indicated food programs no longer existed. The women have had to respond to the changing circumstances and to take on men's breadwinner responsibilities.

The women and girls have no choice but need to venture outside the relative safety area of IDP camps to collect water or to look for firewood to use as cooking fuel on a daily basis. According to one of the interviewees, "collecting firewood for cooking or for selling in the near market, we may be exposing ourselves to greater risk but this the only job we have; we've nothing else to do?" Another affirmed, "even if we know the enemies were there, we go back to the same place again and again to collect firewood". The dangers of venturing out were echoed by others. "I myself have been beaten several times but I count myself fortunate because I haven't sustained serious injuries like some other women," recounted another interviewee. "Looking at current state we just wouldn't know what the future store for us … it's best to keep our mental prepared for any circumstances," and "we always hope for the best but life taught us to be prepared for the worst" indicated two other interviewees.

During the interviews, displaced women indicated that it is the responsibility of women to find a work balance between home-based activities and paid work. Therefore, women in IDP camps find themselves in continuous challenges. Their restricted time availability due to care responsibilities meant they had to seek jobs that offer flexible arrangements. Reflecting on their situation, a group of women at the Abu-Shook IDP camp expressed: "we deny and suppress our pain and cope with suffering rather than embrace it". Another one added "we keep moving forward, opening new doors, and doing new things". "Only God knows how we survive with that sum". "We'd really hope things change after one year to another but all we got is nothing". "Our current one is unrealistic and we're hoping for a new tenure". "We always exercise prudence in our expenditure".

Women in IDP camps have been trapped in informal and low-income work. Informal jobs offer a degree of flexibility to women who have to remain near their home because of care responsibilities. One of the women stated, "you can't be replaced in your family, but you can be replaced on a job, isn't it?" One puts it even more forcefully: "In whatever way I can, I have the RESPONSIBILITY, the DUTY to care for my family". At the end of each interview, I complimented the women. This meant a lot to them, especially given their daily and frequently unacknowledged struggles. Almost all interviewed women revealed that they enjoyed the interview process, and felt they had been able to contribute freely their own perspective about caregiving during and after the war.

Focus groups: giving voice to the previously silent

If individual interviews provide poignant insights into women's lives, focus groups discussions enabled them to share their intersecting perceptions. Kruger and Casey (2009) define focus groups as carefully planned discussions to obtain perceptions on a defined area of interest in a permissive, non-threatening environment. In the different focus groups, the women voiced almost without exception that they had knowledge that was valuable for the reconstruction of their societies. They argued, "all efforts that we have done day after day to ensure life means that we are prepared enough to face reconstruction, so we know what is needed, and thus we should be given a leading role in peace building". Women explained that they know what is needed, "what makes us stronger is our knowledge of how to push forward through difficult situations, like war; we don't rely on others to care for our families". But women complained that "displaced women are largely missing in peacebuilding processes, they massive dependence on external expertise".

Several factors bind displaced people together such as their shared experiences of war and after war. The IDPs come from varied tribal groups and live peacefully and in social harmony. One cannot recognize who are Fur or Zaghawa or *abbala* and *baggara* from the Rizeigat and Missiriya group. I asked the women during focus group discussions, "how about the strength and resilience of IDPs over a lifetime, and building a new, hoping for an even better future?" Women revealed that care practices increasingly bring people affected by war together as they share the living spaces in the camps. Participants in the focus groups indicated that women more than men have the ability to build "resilient communities through their role as caregivers".

Women as caregivers/providers for their family/community contribute to fostering an ethics of tolerance and a spirit of diversity in IDP camps. Women indicated "caring creates an ethical culture, it is moral obligation to show people care … it is a question of respect, responsibility, and trust… these are moral values that are key building blocks to peaceful existence with others". Intermarriage between ethnic groups becomes an important way to minimize violent revenge and it helps to establish family ties between the conflicting ethnic groups. This constitutes a form of social security and stability for IDP communities at large. They succeed in bonding different tribes that previously fought each other. This is a very popular practice in the IDP community for peacekeeping and creates cohesion, and thus women play key roles in setting up resilient communities. As they stated "we always inspire and motivate our people to work together and give their very best".

The caregivers' stories: complexity of the wartime reality

Feminist work is concerned with making women's lives visible (Kennedy et al., 2005). Storytelling offered the Darfurian women who participated in my research the opportunity to communicate their varied experiences, intimate thoughts, and painful feelings verbally. Their stories highlight the unspeakable violence, deep sorrow, and frustration as well as social responsibility, strength, and moments of joys embedded in the landscape of caregiving inhabited by these women.

Most described their painful journey from their homes to the IDP camps through the years of war. They had witnessed the murder of their beloved ones, the destruction of their property, and injury of close family members. They feel their life and those of their family members are in danger, and they constantly have to find new ways of protecting themselves. One woman narrated: "We have saved lives of many during war where there is no any external help". "There was a sexual abuse … those who were raped need special care … It is hard work, it is emotional work, it's physically draining". Another recounted:

> I fled the situation empty handed; I only keep thinking where will we go? How will I find foods for my children? How will I keep my children safe? I must take personal responsibility towards my family. I can't change the circumstances I need to stay strong for those who need my care most.

These concerns have widespread resonance among displaced Darfurian women. A 40-year-old widow and a mother of four tells her story, "I was tired, frustrated and of course in pain, but I didn't blink an eye for my dependents, I kept myself awake most of the time". She paused and took a deep breath before going on:

> I have almost no means of making a living; I just was wondering how to keep my family fed and safe. So many times, I wanted to scream and cry, but it makes it all worthwhile knowing I wasn't the only one that struggled for managing with this situation.

A woman who is now living in the Zamzam camp added: "We need to be strong to face challenges that are put in our path and time will heal all sorrows… We also need to learn to cope and care the best we can…" Zenab who is now working as a teacher:

> have better compassion and empathy for many orphans in the camp who are in need of my help most; I help elderly people in their daily lives to see their happy faces; it is the most rewarding feeling in this life … It gave me a deep sense of satisfaction.

Storytelling gives a much better understanding of the complexity of caregiving arrangements during difficult situations. The quotations above show how women experienced and addressed caregiving, and how they understand their responsibility as pivotal for the survival and welfare of their family and community.

No one can teach a lesson more than one who has experienced it

Ultimately, the women tell their stories not only to convey their experiences, but to also seek to transform them. Amnah, a teacher, resident of one of the IDP camps, a passionate activist for women rights and participant in the study, did not only speak about the reality of women as caregivers, but also about the necessity for change and justice. She said:

> Women traditional feminine work continues utilizing their care giving abilities, and extends it into issues of inhumane practices. During wartime, if it were not for women, the suffering of many (especially children and elderly, wounded and disabled) would have been much greater as the conflict went on. Women in every IDP camp are now seeking justice and fairness, speaking out against violence, seeking redress even against their own father/husband, to end their subordination and rights violations. It is an example of determination to create reform and live as a dignified woman irrespective of whatever deterrents abound. But there is hope, because after war we have woken up from a very deep slumber.
>
> More displaced women are organizing to stand for their individual rights and use the opportunity. We have come to recognize that we all as displaced women are living in IDP camps together and we have a role to play in reforming the tradition that disproportionate burden of unpaid care-work falls on our shoulders, which has informed our associations for change. We have seen how our caregiving works during and after war are undervalued and thereby cast as low status work in peace processes. We have seen how our caregiving works during and after war are undervalued and thereby cast as low status work in peace processes. We must continue to move towards reduction and redistribution of care-work and increasing representation of women equal to men in peacebuilding and reconstruction processes. So we can begin to live well both in young and old age, and equally with men, and transmit proper legacies to our children, even those yet unborn.
>
> What we need to do? We must speak up, and continue to do so as caregivers/providers, have confidence in our abilities and fight for ourselves and our rights, shake unequal and stereotypic care-work systems and remove social inequalities. We need to find ways to involve men in care-work and domestic activities to move to a society of equality and partnership. Men and boys are definitely a vital part of the equation in our need to assert our dignity and respect as human beings with equal rights as them. Every woman desires her husband to share with

her household care burden for the survival of the family, since family and marriage is about partnership is a workable approach to bridging the household division of labor. As caregivers we must have an appropriate self-definition of who we are, believe in ourselves, forge ahead with resistance against any opposition.

Identifying new ways of addressing caregiving and worrying less about being discriminating because we during and after war arrange care-work that contribute to the survival and wellbeing of our family and community. We need to believe in ourselves with more confidence of ourselves to actually use our everyday life experience during war and in IDP camps and our power of desiring to be included in peace making and building in a much more constructive way. Realizing ourselves and growing in confidence will give us a voice as policymakers get attracted to listen to us. Which is why, today the question of women human rights is on the agenda in peace. This is great and exactly what women's struggle has been able to achieve. The question comes-up here do women knowledgeable enough to willingly determine to give an end to violations of their rights and not willing to compromise or let any violations continue?

The violence being one of the most important issues of our lives during and after war, it is vital to women's health and human rights, making injustice visible through the use of whatever means. Because we no longer can be kept in silence, but will take decisive actions that is going to make a lot of progress for all of us as caregivers. All of these displaced women including myself agree that women are more and more understanding of the tremendous importance of coming together, supporting each other and strengthening ourselves and our community. We must have confidence in ourselves, see ourselves as peacemakers/builders, and empower everyone around us- our girls, boys, and men too, so we can change the current situation.

We can network between each other as displaced women everywhere in Darfur to expand the indigenous knowledge we have, and create opportunities of influence amongst one another no matter the distance or difference. So also is education for women, a very important tool, whether formal or informal women and girls need to be literate. The traditional and cultural norms and expectations on women to get married and by housewives as well as more practical obstacles such as the obligation to do housework, work in the field or take care of younger siblings stand on the way of women's education. We need to push and keep pushing towards changing the social status and an outdated stereotype, sharing equally with men care-work activities and greater participation in peacebuilding and reconstruction processes.

Conclusion

Feminists have been concerned with the ways in which care is often feminized work. The Darfur case study shows that during and after war, women carry the greatest burden of caring for their families and communities, even as some of them deal with the traumatic impacts of the war. Socialized into traditional gender roles, many Darfurian women have adjusted more readily to stressful situations, often more successfully than men, and have been the most effective caregivers to the most affected victims. However, wartime caregiving arrangements should not be regarded as a "costless and effective response to such humanitarian crisis – using available resources and skills". This undervaluation of care work and lack of consideration of women's needs and interests, at macro- and micro-levels, have undermined their enjoyment of benefits during post-conflict reconstruction.

The experiences and work of caregiver, therefore, need to be factored in humanitarian responses, and their interests and perspectives should be integrated in security programs and reconstruction efforts. Caregiving must be an essential part of the discussion and formulation of policies to promote gender equity and achieve women's empowerment. The integration of women's interests, participation, and perspectives into conflict resolution and peacebuilding processes can make these more comprehensive, broader in scope, and more inclusive. This also fosters more constructive uses of power, security, and peace in post-war environments.

References

Abdul-Jalil, M. A. (2008). Nomad-sedentary Relations and the Question of Land Rights in Darfur: From Complementarity to Conflict. *In Richard Rottenburg (Hg.) Nomadic-sedentary Relations and Failing State Institutions in Darfur and Kordofan (Sudan).* Halle, Germany: Martin Luther University of Halle-Wittenberg, pp. 1–24.

Ali, H.E. (2014) Darfur's *Political Economy: A Quest for Development.* First Edition, London: Routledge, Taylor & Francis Ltd.

Bakker, I. (2007). Social Reproduction and the Constitution of a Gendered Political Economy. *New Political Economy*, 12, 4, 541–556.

Budlender, D. (2010). *Time Use Studies and Unpaid Care Work.* 1st Edition, New York: Routledge.

Central Bureau of Statistics. (2017). http://www.cbs.gov.sd/.

Ferrant, G., Pesando, L. M. and Nowacka, K. (2014). Unpaid Care Work: The Missing Link in the Analysis of Gender Gaps in Labor Outcomes. OECD Development Centre, Retrieved April 24, 2016.

Friedman, M. (2009). Review of Daniel Engster, the Heart of Justice: Care Ethics and Political Theory. *The Philosophical Review***,** 118, 2, 256–258.

Gilligan, C. (1982). *In a Different Voice: Psychological Theory and Women's Development.* Cambridge, MA: Harvard University Press.

Kennedy, M., Lubelska, C. and Walsh, V. (2005). *Making Connections: Women's Studies, Women's Movements, Women's Lives.* London and Washington, DC: Taylor & Francis – e-Library.

Kleinman, A. (2009). Caregiving: The Odyssey of Becoming More Human. *The Lancet*, 373, 9660, 292–293.

Krueger, R. A. and Casey, M. A. (2009).*Focus Groups: A Practical Guide for Applied Research.* 4th Edition, Thousand Oaks, CA: Sage.

Lawson, V. (2007). Geographies of Care and Responsibility. *Annals of the Association of American Geographers*, 97, 1, 1–11.

Liera, A. and Saraceno, C. (2002). Care: Actors, Relationships and Context. In B. Hobson, J. Lewis and B. Siim (eds.) *Contested Concepts in Gender and Social Politics*. Cheltenham: Edward Elgar.

Noddings, N. (2002). *Starting at Home: Caring and Social Policy*. Berkeley: University of California Press.

Orozco, A. P. and Leiras, A. A. (2017). *Why We Care about Care, An Essay Collection on Care Economy*. The UN Women Training Centre Booklet – Available at the Following Link https://trainingcentre.unwomen.org/mod/data/view.php?d=1&rid=5340.

Peterson, V. S. (2002). Rewriting (Global) Political Economy as Reproductive, Productive, and Virtual (Foucauldian) Economies. *International Feminist Journal of Politics*, 4, 1, 1–30.

Pini, B. (2002). Focus Groups, Feminist Research and Farm Women: Opportunities for Empowerment in Rural Social Research. *Journal of Rural Studies*, 18, 339–351.

Sevenhuijsen, S., Bozalek, V., Gouws, A. and Minnar-McDonald, M. (2003). South African Social Welfare Policy: An Analysis Using the Ethic of Care. *Critical Social Policy*, 23, 3, 299–321.

Tronto, J. (1993). *Moral Boundaries: A Political Argument for an Ethic of Care*. New York: Routledge.

Tronto, J. T (2013). *Caring Democracy: Markets, Equality and Justice*. New York and London: New York University Press.

9 Research in a conflict and peacebuilding context

Narrations from fieldwork in Nigerian violent theatres

Jimam T. Lar

Introduction

This chapter reflects on my experience of conducting fieldwork within a conflict and peacebuilding context. It reflects in particular on how disciplinary constraints on one hand and the socio-political context on the other influence methodological decisions. The reflections are drawn from fieldwork experience in northern Nigeria, first as part of my doctoral research and subsequently for post-doctoral research. The fieldwork took place in three main phases: December 2014–January 2015 in Jos, Plateau state; August–September 2016 in Tafawa Balewa, Bauchi state; and August–September 2017 at the Kaduna archives and Katsina, Katsina state. Though my core research was not focused on the conflict, understanding conflict and its varied dynamics in my research contexts was a critical methodological first step. The specificity of conflict as a research object inevitably has a way of impacting the researcher and his or her study.

My reflections are anchored on five key elements. First, I start with a discussion of my intellectual journey into conflict and peacebuilding research, particularly the thorny matter of disciplinary approaches and methodological socialization. Second, I explore how contemporary realities impact historical narratives. Historical narratives are not free of interest and perspective; they often provide substantive justifications for political and social movements in the present. How does a researcher approach the research terrain, particularly where historical narratives are contested? Third, I reflect on how archival research can be affected by violent contemporary conflict. Since some cases of violent sectarian conflicts are woven around territorial claims to land and autochthony, archival documents emerge as important sources of evidence to present before judicial commissions of inquiry which have become a major process of state engineered peacebuilding.

It is not only the state which uses the archive as a mediating instrument, but different ethnic and social groups also go to it in search of evidence, to strengthen their claims, and unfortunately sometimes to vandalize or destroy documents that their perceived opponents may find useful. As I came to find out, this type of practice has a profound impact on the research process.

The fourth point relates to questions of my identities and perceptions of my positionality, and how they situated me within insider-outsider dynamics. Insider-outsider dynamics have been one of the more widely discussed aspects of field research. Being an insider or outsider could be an advantage or disadvantageous, depending on the context. This dynamic becomes even more acute in a conflict context. I realized that getting access was not only challenging in situations where I was regarded as an outsider, but also where I was an insider. The fifth point of reflection is the challenge of observing and witnessing violence as a result of policing practices. How should a researcher approach this problem? Is it an ethical risk, a matter of vital access, or both?

History, peacebuilding, and multidisciplinarity

My academic interests and intellectual endeavor are informed by a multidisciplinary approach. At a very early stage, after I had defined my area of research interest, it became clear to me that issues of security provision, conflict studies, and peacebuilding had been dominated by the social sciences, particularly by political science. I was motivated to explore to the extent of the insights that historical approach could bring to such debates and discourses. Historians in general study a wide range of topics and phenomena, and employ a much wider variety of methods. There are cases where the subject matter of study benefits from engaging with other disciplines while still relying on historical approach as the core methodology. The historical and contemporary salience of the phenomenon of security, violence, and peacebuilding has led me to adopt a multidisciplinary approach that draws from history, anthropology, and the social sciences more broadly.

In addition to interdisciplinarity, my research and scholarship have focused on the history of ordinary people. In recapitulating the intellectual history of modern historical writing, Richard Evans notes that

> History...when it emerged during the professionalization process of the 19th century, was emphatically the political history of the nation-state and its relations with other nation states. The history of high politics and international diplomacy was king. History was made by great men.
> (Evans 2000, p. 161)

In my specific area of research, namely security, violence, and peacebuilding, my studies are first and foremost an account of the history of security, conflict, and peacebuilding from below. They attempt to understand how politics at the top echelons impacts on the provision everyday security and vice versa. As Hobsbawm notes, "we are not merely trying to give it [history] a retrospective political significance which it did not always have, we are trying more generally to explore an unknown dimension of the past" (1997, p. 270). The overall objective thus is beyond understanding micro and macro trajectories as such; rather it is to analyze the interaction between the two. In retrieving the history of ordinary people, different voices emanate; textured by age, class, ethnicity, gender, and status, these voices generate multiple and competing historical perspectives. While these competing perspectives may exacerbate ongoing conflicts, they also do contain possibilities for policy recommendations and opportunities for peacebuilding.

Contextualizing the present: the past as scarce resource

Arjun Appadurai has shown how historical narratives can be mobilized for contemporary political and cultural projects in the present (1981). The past, he argues, becomes a "scarce resource" within contexts of contestation, because cultural understandings of history may differ, records may be thin on the ground, and one particular interpretation of the past or of oral history may be challenged by another. The right to have traditional-cultural and political suzerainty over Jos, the Plateau State capital or Tafawa Balewa town – the headquarters of Tafawa Balewa Local Government Area of Bauchi state, northeast Nigeria – for instance, is weaved into competing discourses of autochthony, belonging, and ownership. Those invoking rights attached to these notions put forward varied accounts of their "indigenousness" that are often conflicting and contradictory, and above all contested (Smyth 2005, p. 49).

Autochthony as "belonging", according to Geschiere, can have a varied meaning among different groups, even when these groups reside within the same given territory. Fourchard and Segatti suggest a departure from a unifying meaning of all xenophobic tendencies. The conceptualization offered considers synonymous notions "as autochthony, nativism, and indigeneity as local concepts used by actors in situations of xenophobia" (Fourchard and Segatti 2015). Autochthony is therefore situated within a specific social context of discourse and practice. For some groups, the claim to autochthony is based on early settlement, for others they are autochthonous because of conquest, and yet other groups claim to be autochthonous on the basis of having lived in an area for a given period of time over several generations (Geschiere 2009).

Two critical examples of such contestation from the field are the cases of "indigene settler violence on the Jos Plateau, central Nigeria" and the cases of collective violence in Tafawa Balewa in Bauchi state (Odoemene 2008; Nwaka 2016). The violence in Jos has been seen as linked to claims of ownership and the struggle over who is an indigene and who is a settler. In this context, contemporary politics has had a profound influence on how ethno-religious groups' histories are narrated. The conflict in Tafawa Balewa in Bauchi state also poses an interesting set of questions, with issues of indigeneity arising in the conflict between the Sayawa and Fulani in Tafawa Balewa, but also because the recurrent episodes of collective violence on the Jos Plateau have been keenly felt in neighboring Bauchi State. Indigeneity discourses in Bauchi have been influenced by the Plateau conflicts and vice versa. In both contexts, the historical narrative is weaponized, and indigeneity and autochthony are used as tools of exclusion.

Communities and groups construct collective accounts to buttress and provide legitimacy to their claims. As I have realized from my own research experience, historical narratives are not free of interest and perspective; they provide a historical justification for political and social movements in the present (Evans 2000, p. 149). The writing of history or historiography illuminates two periods: the one it is written about and the one it is written in (Bragg 2000). So it is possible for informants, communities, and groups to situate the basis of their contemporary claims on historical legitimation. This, therefore, necessitates the researcher to approach the study of the history of violent conflict and peacebuilding from a vantage point of the present; this is the language of entry, acceptance, and access.

Starting with the history, which is what researchers actually seek, has a tendency to put off the informant in a study. Fred Cooper asserts that trying to illuminate present issues is a fine motivation for exploring the past; however, as one looks backwards, there is a risk of anachronism: confusing the analytical categories of the present with the "native" categories of the past (Cooper 2005, p. 18). Of course, there is a risk of anachronism, particularly when researchers approach the past with the logics of the present to render reliable and trustworthy historical accounts, it is critical that researchers understand the historical past in its own terms and context. This applies to the social, political, and cultural contexts of the specific periods that are studied.

Historians are methodologically socialized to be cautious about how they use, interpret, and give analytical value to their evidence. The first and primary obligation of the historian is to establish that an event did occur, and then analyze what it might have meant to different protagonists and with what specific consequences. The aim of this forensic procedure is the reconstruction of historical realities. Historical knowledge about facts is therefore both desirable and attainable. However, when we reflect on our data and in the process of analyzing and understanding what it is telling us, critical insights may emerge.

Insider, outsider, and dynamics of field research

During my fieldwork, I had to confront questions about my identity(ies) and perceptions of my positionality, and how they are situated within insider-outsider dynamics. Writing history with the motivation that such a study will provide understanding on the present period, which the researcher lives in, raises questions of positionality. Eminent Nigerian historian Yusufu Bala Usman shows the importance of this point. According to Usman:

> since the purpose of historical study is to grasp the historical process for the purpose of influencing it then the only correct conceptual framework, for any epoch is that which provides a basis for the practical

> determination of the direction of (agency) in that epoch...every view you have of the past has specific implications regarding what you should do now.
>
> (Usman 2006, p. 22)

I have been socialized by the collective experience of non-Muslim ethnic groups of central Nigeria, and this experience is informed by a history of hostility between the non-Muslim groups and the Hausa-Fulani emirates.

While there is a history of trade and cordial relations, the predominant dynamic of the relationship between the Hausa-Fulani emirates of Northern Nigeria and the non-Muslim communities of central Nigeria before the arrival of the British is characterized by denigration of the non-Muslim groups by the Hausa-Fulani, which is reciprocated with distrust and suspicion. To understand this dialectic, there is need to briefly recapitulate the history of the relationship, and highlight from a central Nigerian perspective some of the main factors that have defined the representation of Hausa-Fulani in the consciousness of the peoples of central Nigeria. Two key points will suffice. First, the wars and experiences were occasioned by the extension of the Sokoto Jihad to the non-Muslim communities. Second, there were long-standing derogatory references to these areas as a backwater used for the purposes of slave raiding. In the wake of the Sokoto Jihad, armies from Bauchi and the *Zazzau* (Zaria) emirates endeavored to conquer the Plateau polities, and while they did make substantial inroads and penetrated the area, they did not succeed in conquering the people (Mangvwat 2013, p. 112).

The British who invaded the Plateau communities by the early 20th century experienced the same resistance that jihadists had encountered earlier, though with a different consequence, owing largely to superior weaponry. British conquest would further extend the Hausa political, social, and cultural influence into the Middle Belt area. Prior to British colonization, institutionalized Hausa influence in the Middle belt had been limited to a few emirates, established following the Sokoto Jihad in the early 19th century within the Middle Belt area.[1] Ochuno has termed Hausa influence as "scattered systems" of emirate control that nominally oversaw some non-Muslim peoples while pragmatically ignoring others. Memories of these experiences are still alive across the present Plateau State. British colonization introduced an administrative policy that placed non-Muslims under the control of emirates. This misreading of precolonial relations between the emirates (like Bauchi and Zaria) and the non-Muslim groups on the Plateau highlands and the Jema'a area in Kaduna state was continually resisted throughout the period of British colonization.

The culmination of this resistance was the rise in Middle Belt consciousness (Ochuno 2014, pp. 77–85). This explains why even contemporary sectarian violence in Plateau State occasioned by proximate identity politics is easily associated within the consciousness of the hitherto non-Muslim groups with Hausa-Fulani historical hegemonic tendencies. As Smyth has argued, violent conflict tends to consolidate and reinforce such boundaries as the wider society polarizes in response to violent attacks and counter-attacks (Smyth 2005). In research circumstances like these, personal reflexivity is imperative for the process of history writing. Hence, Usman extrapolates the point further: "Once you are aware of your own historicity, then you will realize the historicity of the concepts that you use, and you do not sit down and accept them as if these have come down to you from heaven" (Usman 2006, p. 92). In other words, recognizing one's own biases, positions, and subjectivities informed by one's historical experience is a critical starting point. Furthermore, it is important to recognize our positionality in the process of fieldwork, interpreting data, and writing. One has to recognize assumptions that are drawn due to the influence of socialization and to rigorously resist the temptation to let these assumptions direct our inquiry. Consequently, I am attempting here to study the history of security provision, conflict, and peacebuilding beyond the received knowledge in scholarly work and in public discourse in Nigerian society. Earlier assumptions and perceptions informed by my socialization are therefore relatively mitigated by the information that emerges from the research process.

Beyond the conflict context, insider-outsider dynamics of a research context have been one of the more discussed aspects of field research (Merton 1972; Hermann 2001; Smyth 2005). It raises questions like who is an insider and who is an outsider? This becomes even more acute in a conflict context, as I realized getting access was not just challenging where I was regarded as an outsider but also where I was seen as an insider. In the context of my research, insiders and outsiders are understood around the primary markers of religion and ethnicity. Beyond this, we also find professional communities. Following the several episodes of violence, Jos became segregated along religious lines (Krause 2011, pp. 44–46). Violently divided societies often are characterized by geographical and ideological segregation between conflict zones and safe territory, with researchers taking the risk to venture into conflict zones for the purpose of doing fieldwork (Smyth 2005, p. 18). Therefore, as a Christian resident of Jos, I had to consider if it was safe going into the Muslim quarters of the city for research purposes. Since I desired to meet and talk to Muslim vigilante members and groups meant that I had to try, but it required caution and good preparation. The challenge was not only that of physical safety, it was also about access – so that people could be in a position where they trust me and could feel free to talk to me. In such a context of conflict and suspicion, it is difficult for a researcher to convince informants that he or she comes without an opposing religious or political predisposition.

Contestations, conflicts, and the archive

I conducted archival research after I had spoken to key informants. Therefore, I already had some themes, concepts, and perspectives I was considering. In the archives, I came across opinions and views that supported some of my perspectives, but altered a number of others. With archival documents, it is not just the text that one considers; also important is an understanding of who wrote it and in what given context – this could be social, political, or legal. In this sense the researcher finds himself or herself in conversation with the archival source; accordingly one cannot claim to "recover a single, unalterably 'true' meaning of a dispatch simply by reading it; on the other hand, we cannot impose any meaning we wish to on such a text either" (Evans 2000, p. 106). A document also has meaning in relation to other documents; in this case, the researcher is also interested in what preceded and what follows. Documents are therefore studied as pieces of a puzzle and there are occasions where the puzzle is not completed and the researcher has to reconsider lines of argument. Therefore, when a historian goes to the archives, one is not in search of a toolkit of evidence to answer all questions; rather it is a tedious process of reading heaps of documents and trying to make sense of it all. At the end of the process, the archival material complements the process of reconstructing the historical narrative. This explains why it is most frustrating when documents in the archives are found to be missing, or have been looted, or vandalized.

My research experience has shown that archival research can also be impacted by violent conflict. One may imagine that this may relate to the archives or libraries being destroyed, as was the case in Timbuktu, Mali, in 2013. That being an important risk, in my case, however, I refer to a different kind of vandalization. Owing to the fact that some cases of violent sectarian conflicts are woven around territorial claims to land and autochthony, colonial documents kept at the archives emerge as important sources of evidence to present before judicial commissions of inquiry. One of the government's responses to recurring episodes of violence has invariably been to institute commissions of inquiry, often headed by legal luminaries, with the mandate of establishing the causes of specific episodes of violence, identifying the key actors, and proposing remedial measures to prevent future violence (Suleiman 2015).

Groups go to the archives in search of evidence to strengthen their claims, but unfortunately to also vandalize potential evidence that perceived opponents might find useful. At the Kaduna Archives, I found evidence of visits by ethnic organizations in visitors' register. This was around the period when the organizations were preparing their memorandums to submit to the Justice Bola Ajibola Commission of Inquiry that followed the 2010 electoral violence in Jos. For instance, at the Kaduna Archives, missing documents include memos on chiefs and their loyalty or the lack of it, minutes of meetings, political activities in Jos in the 1950s, and proposals for reorganizing divisional and native authority boundaries. Adoyi, referring to such practice as the malaise of pilfering through stealing of critical and sensitive archival documents, identifies some other objectives like attempting to influence chieftaincy disputes, land cases, and/or certain damaging intelligence reports compiled about areas as reasons why archival documents are taken (Onoja

2010, p. 109). While we may not be able to determine the identity of the people responsible for these acts, one may point to interesting coincidence that documents missing are related to such areas engulfed in sectarian conflict.

Witnessing violence

When one is researching contemporary vigilante practice, as I did in parts of Plateau state in 2012 and 2014 and in Zamfara and Katsina states in 2017, a key source of data is observation. While interviews can provide insights into the structures, functions, and individual motivations for joining such vigilante groups, observing everyday practice for a sustained period complements and sometimes contradicts the oral information provided by informants. Getting access to go around with vigilante groups as they go about their normal daily routines can constitute a challenge on two levels. Vigilante groups work during the day – for instance patrolling local markets – but the main function of guarding is actually at night. In taking the decision to request permission to go out on night patrol, one has to weigh the risks involved; questions of (the researcher's) safety are to be considered. In other words, when making such a decision, you take the same risk which the vigilantes take. Vigilantes are never certain of the situation that awaits them every night they go out, how armed or dangerous a potential offender or suspect may turn out to be. When going out on patrol with a Vigilante Group of Nigeria (VGN) member as a researcher, all these considerations are also important.

The second level entails the extent to which the researcher's presence can impact or influence practice. One of the key objectives of observation is to be as inconspicuous as possible; in other words, to what extent is one's research partners carrying out their normal daily routine, or is it a case of performance? In my case, the question that emerged was how should I deal with instances of "violent interrogation" – this to an extent is an ethical dilemma. Should I decline to conduct observational research of vigilante or policing practice, and simply rely on what I am told? Should I speak out when I observe that a line is perhaps being crossed? Or should I merely observe and basically describe and analyze the practice as experienced?

In this case, I refer to Lars Buur, confronted with a similar dilemma. Buur decided to witness and analyze vigilante activities. He notes that:

> ...just because one actually experiences something, (violence) this does not mean one necessarily endorses what one sees and experiences. Sometimes the best way of speaking up against something is by describing it so that others can make their own judgements.
>
> (Buur 2010, p. 47).

Baz Lecocq laments how "positive, and especially negative emotions, within the field at large, and its effects on analysis and writing are scarcely dealt with" (2002, p. 274). Furthermore, commenting on what he described as the open reluctance and outright denial of the "pain of fieldwork" in favor of the "fieldwork holiday myth", Lecocq contends, "not accounting for all of one's attitudes in and towards the field in one's professional life and writing is to some extent cheating both the audience and oneself. It is simply denying one's own humanity" (Ibid).

It is worth noting that I was working in areas I had lived; I was thus familiar with the general context. The more serious challenge in regards to access during fieldwork was gaining access to the Muslim side and overcoming any suspicion they may have felt towards me. In approaching fieldwork in conflict and violent context, it is a given that as a researcher one should not precipitate further violence. That said, there is definitely an objective where possible to provide constructive recommendations towards peacebuilding, or as was the case with my research that attempted to explore and to understand local actors' efforts to resolve conflicts from below (Lar 2017; Mustapha, Higazi, Lar and Chromy 2018).

One of the major points I came away with was how thin the line is between negotiating these challenges and being stuck in the field bereft of any idea of what to do. In other words, despite all these challenges, I would argue that I generally managed to achieve my research objectives, while still noting that it could have easily gone wrong.

Conclusion

In conducting research, particularly conflict and peacebuilding research, there are particular methodological pitfalls and issues researchers have to consider and come to terms with. They need to understand the context of conflict and its various dynamics, even when their research is not necessarily focused on the conflict. As discussed in this chapter, even archival and historical research can be impacted by violent conflict. Contemporary realities impact on historical evidence, narratives, and interpretations. We should always ask whether our training for field research in conflict zones or conflict context is adequate or still lacking in the manner we are prepared for the realities of such contexts. Research projects generally have a possible means:

> of exploring the processuality of research in both the historical and the social sense. This would entail scrutinising the various research practices in their textbook sequence, from development of questions and hypothesis or given perspectives found in the literature, through the various types and situations of data collection, to the evaluation and write up phase.
> (von Oppen 2003, p. 57)

As a rule, and from my experience, fieldwork is dynamic and flexible. Even with the best pre-planning and structure, there are so many unexpected occurrences in the field which can alter and change research in a conflict and peacebuilding context.

Note

1 Examples of such emirates in the Plateau Province were Wase and Kanam.

References

Appadurai, Arjun. (1981) "The Past as a Scarce Resource", *Man*, Vol. 16, No. 2, pp. 201–219.

Bragg, Melvin. (2000) "In Our Time", *BBC Radio* 4. March 30th.

Buur, Lars. (2010) "Domesticating Sovereigns: The Changing Nature of Vigilante Groups in South Africa", in Thomas Kirsch and Tilo Gratz, eds. *Domesticating Vigilantism in Africa*, Suffolk and Rochester: James Currey, pp. 26–50.

Cooper, Fredrick. (2005) *Colonialism in Question: Theory, Knowledge, and History*, Berkeley and Los Angeles: University of California Press.

Evans, Richard. (2000) *In Defence of History*, New York: Norton.

Fourchard, Laurent and Segatti, Aurelia. (2015) "Introduction of Xenophobia and Citizenship: The Everyday Politics of Exclusion and Inclusion", *Africa*, Vol. 85, No. 1, pp. 2–12.

Geschiere, Peter. (2009) *Pearls of Belonging: Autochthony, Citizenship, and Exclusion in Africa and Europe*, Chicago, IL: University of Chicago Press.

Hermann, Tamar. (2001) "The Impermeable Identity Wall: The Study of Violent Conflicts by 'Insiders' and 'Outsiders'", in Marie Smyth and Gillian Robinson, eds., *Researching Violently Divided Societies: Ethical and Methodological Issues*, London: Pluto Press and Tokyo: United Nations University Press, pp. 77–91.

Hobsbawm, Eric. (1997) *On History*, London: Abacus.

Krause, Jana. (2011) *A Deadly Cycle: Ethno-Religious Conflict in Jos, Plateau State, Nigeria*, Working Paper: Geneva Declaration Secretariat, Geneva.

Lar, Jimam. (2017) "Local Actors, Conflict Management, and Peacebuilding in Central Nigeria: Insights and Policy Implications", APN Briefing Note Number 9, New York: African Peacebuilding Network (APN) – Social Science Research Council (SSRC).

Lecocq, Baz. (2002) "Fieldwork Ain't Always Fun: Public and Hidden Discourses on Fieldwork", in *History in Africa*, Vol. 29, pp. 273–282.

Mangvwat, Monday. (2013) *A History of Class Formation in the Plateau Province of Nigeria, 1902–1960, The Genesis of a Ruling Class*, Durham: CAP African World Series.

Merton, Robert K. (1972) "Insiders and Outsiders: A Chapter in the Sociology of Knowledge",

Mustapha, Abdul-Raufu, Higazi, Adam, Lar, Jimam, and Chromy, Karel (2018) "Jos: A Decade of Fear and Violence in Central Nigeria," in Mustapha, A. Raufu and D. Ehrhardt, eds., *Creed and Grievance: Muslims, Christians and Society in Northern Nigeria*, Suffolk: James Currey, Rochester: Boydell and Brewer Ltd, pp. 225–268.

Nwaka, Jacinta Chiamaka. (2016) "Religion, Economy and the Pre-colonial Dimensions of the Jos Conflict", *Afrika Zamani*, No. 24, pp. 31–44.

Ochuno, Moses E. (2014) *Colonialism by Proxy: Hausa Imperial Agents and Middle Belt Consciousness in Nigeria*. Bloomington and Indianapolis: Indiana University Press.

Odoemene, Akachi. (2008) "The Contexts of Colonialism and Ethnicity in Indegene-Settler Relations: Comparative Evidence from Social (Dis)order in Two Nigerian Cities", in Bahru Zewde, ed., *Society, State and Identity in African History*, Addis Ababa: Forum for Social Studies, pp. 231–256.

Onoja, Adoyi. (2010) "An Endangered Species? The State of the Archives in the Reconstruction of Nigerian History", in Mike O. Odey, John J. Nengel, and Okpeh O., eds., *Historical Research and Methodology in Africa: Essays in Honour of Professor Charles Creswell Jacobs*, Makurdi: Aboki Publishers, pp. 104–114.

Porter, Elisabeth, Robinson, Gillian, Smyth, Marie, Albrecht Schnabel, and Eghosa, Osaghae. (eds.). (2005) *Researching Conflict in Africa: Insights and Experiences*, Tokyo, New York and Paris: United Nations University Press.

Smyth, Marie. (2005) "Insider-outsider Issues in Researching Violent and Divided Societies", in Elisabeth Porter, Gillian Robinson, Marie Smyth, Albrecht Schnabel, and Eghosa Osaghae, (eds)., *Researching Conflict in Africa: Insights and Experiences*, Tokyo, New York and Paris: United Nations University Press.

Suleiman, Samaila. (2015) "The Nigerian History Machine and the Production of Middle Belt Historiography," Ph.D. Thesis, University of Cape Town, January 2015.

Usman, Yusufu Bala. (2006) "The Assessment of Primary Sources: Heinrich Barth in Katsina, 1851–1854," in George Amale Kwanashie, Alkasum Abba, Saidu Hassan Adamu, and Barira Mohammed, eds., *Beyond Fairy Tales: Selected Historical Writing of Yusufu Bala Usman*, Zaria: Abdullahi Smith Centre for Historical Research, pp. 1–22.

Von Oppen, Achim. (2003) *Bounding Villages: The Enclosure of Locality in Central Africa, 1890s to 1990s*, Berlin: Habilitation Treatise, Humboldt University.

10 Epistemological and methodological considerations in peacebuilding research – experiences from the Borana of Ethiopia

Asebe Regassa Debelo

Peacebuilding research in Africa requires conceptualizing and contextualizing notions of peace, conflict and inter-group relations within cultural, geographical and political settings at local and national and regional levels. However, studies on conflict and peacebuilding in the continent are often built on theories, epistemologies and methodologies developed in Western contexts. On the contrary, among many indigenous communities in Africa, peace is epistemologically conceptualized as harmonious relationship between humans, nonhumans and supernatural power. In such context, peacebuilding is understood as a culturally embedded practice that is performed within broader ontological and epistemological understandings of the people. Building on empirical data from my fieldwork along the Ethiopia-Kenya border, this chapter provides epistemological and methodological insights to peacebuilding research and calls for a paradigm shift towards culture-sensitive approaches. This chapter argues that researching peacebuilding in Africa should consider local cultural settings and epistemological paradigms by carefully designing appropriate methodological approaches.

Introduction

Over the last four years, I have been conducting fieldwork among the Borana Oromo in southern Ethiopia. These experiences have made apparent some methodological challenges social science researchers might encounter while conducting research on conflict and peacebuilding. I was often challenged by the Borana elders when asking them about conflict, peace and initiatives of peacebuilding within the dominant conceptualization of these concepts. Within the mainstream Western notions of peace and conflict studies, while peace is conceptualized as the absence of conflict, it narrowly defines conflict as disputes between individuals, groups or institutions (Zeleza 2008). On the contrary, indigenous communities comprehensively understand peace as harmonious coexistence between humans, nonhumans and the cosmos (Chilisa

2012). Drawing on this latter perspective, I argue that peacebuilding research among societies with intact indigenous cultural practices, values and customs should be grounded within the societies' epistemological and cosmological orientations of their place in nature.

Two specific incidents became turning points to my arguments in this chapter. First, I visited Borana land along the Ethiopia-Kenya border in 2015 to conduct research on cross-border conflict funded by the African Peacebuilding Network. At the time, there was no war between the Borana and their neighbors, but Borana elders described that their land was not at peace. Second, in 2017 during a violent conflict with the Somali, I got a different interpretation of the situation in Borana land, which elders unanimously described as the most peaceful period in their history. This prompted me to raise the following questions: What epistemological orientations inform the Borana to conceptualize peace and conflict in the way they do? How do the Borana people practice peacebuilding? Methodologically, I found myself confronted with issues of positionality, "ethnographic seduction" and silence on the side of the people and also distance from the society's culturally embedded knowledge on my side. This also invokes another question on what methodological considerations should be made while doing research on conflict and peacebuilding among indigenous societies like the Borana Oromo.

It has now become recognized that state-centric, top-down peacebuilding approaches through military interventions have often failed to bring sustainable peace (Autesserre 2010; Hutchinson and Pendle 2015). While scholars, politicians and practitioners focus on issues of conflict and conflict resolution at local, national, regional and international levels:

> at the heart of these challenges lies the unabated aspiration for peace, which implies the necessity of finding a way of living together better in this world of growing complexity and uncertainty that all too often is now witnessing the outbreak of new forms of violence.
>
> (UNESCO 2013, p. 1)

Beyond the aspiration, however, the major challenge remains to understand what peace constitutes in different societies. While the dominant notion of peace as an absence of conflict has been grounded in Western mainstream peace and conflict studies, indigenous societies understand it as firmly grounded in their culture encompassing the spiritual, material, intellectual, emotional and cosmological dimensions (Chilisa 2012). In such contexts, the relational ontology of the people – their connectedness with humans and nonhumans, and their relational axiology, which constitutes value systems, beliefs and traditions – shape the way people perceive, practice, experience and reflect on notions of peace and peacebuilding (Chilisa 2012; Regassa 2016).

In order to critically grasp the culturally embedded perspectives of peace, conflict and peacebuilding among the Borana Oromo, this chapter adopts relational ontology, epistemology and axiology within the paradigm of indigenous research methodologies. In her seminal book, *Decolonizing Methodologies*, the Maori scholar Linda Tuhiwai Smith argues that the dominant Western notions of knowledge production should be deconstructed if at all we have to unveil indigenous peoples' knowledge through participatory research (Smith 1999). Smith maintains that in contrast to the individualized knowledge ownership in Western scientific culture, knowledge is culturally grounded, socially produced, communally owned, relationally understood and transmitted from generation to generation among indigenous societies.

Relational ontology, according to Chilisa (2012), constitutes relations with/between people, environment/land, cosmos (spirituality) and ideas. Further explained, it encompasses human-human relations, human-nature relations and the tripartite relationship between humans, nonhumans and supernatural power. The people's place in nature is not built on notions of dualism, which constructs dichotomized relations between humans and nonhumans; rather, indigenous societies perceive that their culture is conjointly constituted with/in nature (Descola 2013; Regassa et al. 2017). Furthermore, building on relational epistemology that recognizes relational ways of knowing, this chapter situates notions of peace among the Borana as a constituted knowledge positioned in people's relations and interactions with each other, with their environments and with the cosmos.

The Borana and their relations with neighboring groups

The Borana Oromo are a pastoralist community who inhabit the arid areas of southern Ethiopia and parts of northern Kenya. The community's transhuman livelihood, the impacts of climate change and the existence of rival ethnic groups in the area have put the people amidst perpetual conflicts (Oba 2013). As a pastoral society, the Borana move with their livestock in search of pasture and water, and also as a mechanism of natural resource management. As a result, they enter into conflicts with their neighbors, including the Somali, Burji, Guji and Gabra in Ethiopia and on the Kenyan side of the border (Adugna 2010; Bassi 2010). Moreover, their location along the Ethiopia-Kenya border makes the territory contestable, both by the national government and local actors for political and economic reasons. Nevertheless, the Borana and their neighboring communities often resolve disputes and conflicts through indigenous mechanisms without waiting for state intervention (Bassi 2010; Regassa 2016). Their conflict resolution is grounded within the society's indigenous socio-political organization called the Gadaa system.

The Borana are well known for their practice of the Gadaa system – an egalitarian socio-political organization in which power transfers from one generational set (party) to another every eight years through democratic processes (Legesse 1973, 2000). The Gadaa system encourages values of peace, coexistence and togetherness. Both the Gadaa system and its spiritual institution called the Qaalluu serve as foundational institutional frameworks in building peace within the Borana and between them and their neighbors – as it applies to the broader Oromo society (Baxter 1991). Due to its egalitarian political system and because of embracing multidimensional values, practices and knowledge pertaining to human-human relations and human-environment relations, the Gadaa system was registered into the UNESCO world heritage in 2016.[1] This makes the Gadaa system an alternative indigenous and local instrument of peacebuilding practice. Nevertheless, there are emerging competing forms of conflict resolution and peacebuilding practices in the area that often bring indigenous and state institutions into conflict and competition (Regassa 2016).

The Borana share borders with many pastoralist and agro-pastoralist groups, including Guji Oromo, Gabra, Garri-Somali, Rendille, Burji, Konso, Hammer, Tsemai and many others. As a pastoralist community, the Borana enter into conflicts with their neighbors due to different reasons, chief among them being competition over resources, issues of territoriality and culturally and economically grounded practices of cattle rustling (Schlee 2003; Bassi 2010). Although conflict has become a defining feature of pastoralist areas in the horn of Africa (Omeje and Hepner 2013), the Borana's relation with their neighbors has been established on mutual interdependence, respect and reciprocity. To understand the notions of mutual interdependence and reciprocity, it is imperative to further probe into the ecological, cultural, political and socio-economic contexts of pastoralist societies, particularly those in East Africa. Ecological fragility, climate variability and subsequent shortage of water and pasture necessitate the establishment of strong social capital and interdependence among pastoralist societies to be resilient against such environmental stresses (Krätli 2015). Krätli (2015) further argues that climate variability strengthens pastoralist communities' resilience through innovative internal and intra-communal partnerships.

Internally, pastoralist communities have developed, over the years, the knowledge of rangeland management, herding strategies, water resource development and territoriality. In this regard, the Borana are well known for their water resource management and utilization through an indigenous administrative system (Bassi 2005). Accordingly, scarcity is an enabling force for the Borana and other pastoralist communities in prompting them to develop adaptive knowledge in coping with the fragile arid environments. In terms of inter-group relations, climate variability and pastoralist communities' settlement in arid and semi-arid parts of Africa have put them in limbo. On the one hand, according to some scholars on conflict, resource scarcity as a result of climate change and herd population invokes inter-group conflicts among pastoralist communities (Hagmann 2005). According to this perspective, individuals, groups, multinational corporations and states orchestrate inter-group conflicts in pastoralist areas, either in an attempt to access resources in territories controlled by other groups or in response to restriction from other groups in their pursuit of accessing pasture or water points (Homer-Dixon 1994). Scarcity theorists like Homer-Dixon try to link inter-group conflicts in developing countries to increasing predation over scarce resources “like cropland, fresh water, and forests, leading to their degradation, depletion and unequal distribution” (Bhavnani 2009:66).

On the other hand, climate variability, ecological fragility and vulnerability to state and non-state actors predatory practices over their resources become incentive to building strong inter-group relations through values of respect, reciprocity and mutual interdependence. In the case of the Borana and their neighbors, for example, inter-group relations are built on intermarriage, joint ritual participation, reciprocal resource utilization and alliances during warfare with other groups. Such relationship goes beyond national borders in which the Borana extend their partnership with their kin members and other groups on the Kenya side of the Ethiopia-Kenya border (Schlee 2003; Regassa 2016). Therefore, any research on conflict and peacebuilding among pastoralist communities should take into consideration the agency and resourcefulness of the people in contextually responding to internal and external forces.

Fragile environment and its role in inter-group conflict and peacebuilding

The Borana inhabit one of the most fragile and arid environments in the horn of Africa with very limited rainfall and frequent droughts. According to Borbor Bule, a Borana elder, his people face frequent droughts and shortage of water and pasture, which has appeared only in the last three to four decades. According to my discussions with one of the top officials in Borana zone in February 2018, conflict as a result of resource scarcity, which is in turn caused by drought, has become one of the major challenges to the Borana pastoral community. People must travel a long distance to get water and grass for their livestock. The Eela and Tuula (indigenously developed and managed water points) are now drying out. While the youth take their herds far away, they cross into other territories or others cross into Borana territory. This has tended to cause skirmishes that sometimes escalate into inter-group conflict. Recently, however, the conflict has started to take on a different dimension. Many different actors have developed interest in Borana land. The Somali continuously expand to Borana territory. The federal government also has strong interest, particularly on the border area with Kenya. Other actors such as contrabandists also instigate conflict between groups. All these forces have placed the Borana zone under serious challenge from natural and man-made causes (anonymous informant, Yaballo, February 2018).

The above description by the official about the conflict dynamics in the Borana zone hints upon the interplay between climatic variability and emerging interests from different actors in the region. While drought and its consequences are part of the broader climatic change in the region, Somali expansionism and the federal government’s interest on the Ethiopia-Kenya border, and its eventual backing of the Somali in their territorial claim in

Borana land, has further complicated the conflict theatre (Tache and Oba 2009; Regassa 2016). Scholars have already documented the historical process of the Somali expansion to Borana land backed by the successive Ethiopian regimes since the late 19th century, and the process is still ongoing (Bassi 2010).

To understand the salience of climate variability in building local people's resilience and inter-group cohesion, it is imperative to briefly describe the pastoralist way of life. Most pastoralists in East Africa practice transhumant way of life by seasonally moving with their livestock in search of pasture and water (Omeje and Hepner 2013). In contest to the dominant discourse that downplays their knowledge of territoriality, pastoralists are highly connected to place and through their transhumant movements, they establish knowledge and connectedness to territorial spaces. The Borana, for example, have established property regimes ranging from communal to clan-based resource ownership and utilization. While territoriality, as a practice of territorial possession, knowledge and control, makes groups protect their own territory and resources bounded within it, it also opens the contours of their understanding about resource scarcities within their territorial control and the need for depending on other groups for their livestock. Therefore, the Borana value inter-group coexistence and mutual interdependence through values of reciprocity and respect. A Borana elder once succinctly underlined the importance of mutual interdependence and trust as follows:

> Water and grass have become serious problems of our people. Our livestock have to move far away to pass the dry season. Search for grass and water puts our youth in confrontation with herders from other groups. Therefore, our elders have long time ago established principles of mutual interdependence at times of droughts and war to give access to grass and water to one another for their livestock. According to this principle, elders from a clan or a group who faced natural or man made disasters first seek permission from elders where the resources are in good condition. After that, elders would order their youth to allow other herders to herd in their territory. According to this principle, we allow people to our grass and water points because we know that we will also seek their support other time when rain stops on our side. We trust them, and they do also trust us. This is how we live together.
>
> (Informant, Dirre Guyyo, February 12, 2018)

For the Borana, peaceful coexistence with their neighbors is beyond avoidance of conflict. It rather constitutes abiding the principles of reciprocity and respect, and also the notion of giving and receiving refuge from their neighbors at times of disaster. Failure to do so, according to Borana elders, would not only result in lack of reciprocity at times of adversity, but also lead to total break-up of social cohesion, both within the society and between the Borana and their neighbors.

Changing relations with neighbors

Despite a long history of interdependence and coexistence built on values and principles of mutual respect and reciprocity, inter-group relations between the Borana and their neighbors has changed into conflict, particularly in the post 1991 period following the introduction of ethnic federalism in Ethiopia (Abbink 1997; Beriso 2009). Beriso (2009) further argues that the post-1991 political reordering along ethno-linguistic line has reconfigured the alliance style between different groups by changing old enemies into friends but creating new forms of antagonism among old friends. Borana elders attribute such changing relationships with their neighbors, particularly the Garri Somali, to two major factors. First, the Borana perceive that successive regimes in Ethiopia have had a system of differential treatment of citizens in which they backed the Somali in their expansion and control of Borana territories. Some scholars also share the same view, stating that Somali expansionism was backed by the imperial, military and recently the Ethiopian People's Revolutionary Democratic Front (EPRDF) regimes alike. Second, they attribute it to the decline in shared values, institutions and practices of peacebuilding. As Tache and Oba (2009) argue, indigenous institutions that have, for long time, maintained inter-group harmony and coexistence have now been shattered, declined and replaced by state institutions that promote individualized and state-centric conceptualizations of peace and conflict as well as territorialized notions of inter-group relations.

Borana's notions of peace

In order to understand the salience of situating peace research within local contexts, it is important to understand the culturally embedded values and knowledge of peace and conflict among the people. For the Borana, peace is a culturally grounded phenomenon and practice that is nurtured through continuous action and practice. While discussing the notions of peace among the Guji Oromo, a feature that the Borana also share, Baxter (1991) argues that *nagaa* (peace) constitutes ontological, discursive and practical conditions of the relationship within the community, between the community and their neighbors and between them and nonhumans, including the cosmos. Ontologically, the Borana conceptualize peace as a harmonious relationship between humans, nonhumans and the supernatural power. In the tripartite relationship, it entails the smooth communication, flow and coexistence of the natural order without disruption. Discursively, it constitutes the representation of relationships in terms of norms, values and customs in which the youth respect elders, men respect women and vice versa, and the society adhere to ancestral cultural values and rules, and also give reverence to the principles of human-nonhuman relations. In areas of practical engagement, the Borana consider peace as a continuous practice nurtured through rituals, prayers, social mobilization and building institutions and social capital. I will describe each of these categories of peace among the Borana and how they work in building and maintaining peace.

According to Borana elders, peace is broadly defined as harmony between creatures and creator. A Borana elder, Guyo Dida, describes it as follows:

> When uumama [creature] adheres to the seera [law] of the uuma [creator], Borana land becomes peaceful. Humans respect the law of nature, non-humans also abide by the natural laws and coexist with humans. Nagaa [peace] is everything for us. It includes coexistence between humans like our neighbors; it also includes harmony with wild animals and spirits. Peace also goes to harmony with Waaqa [God or supernatural power]. If there is no peace with Waaqa, he might withhold rain and cause drought. Whenever Waaqa is not pleased with the Borana, he invokes conflict.
>
> (Informant, Guyo Dida, February 2018)

A number of interpretations can be made out of the above quotation. Peace depends on humans' reverence to the law of nature that, according to the Borana, prescribes moral and ethical conditions, which humans should follow in order to establish harmonious coexistence in the tripartite relationship. It is not only humans who respect the law of nature. The Borana mythology attests that nonhumans, including livestock, wild animals, plants and spirits, communicate and understand Waaqa's principle in their relations with humans and Waaqa. As a result, they conform to original covenants between the three constituents of nature (humans, nonhumans and Waaqa). The elder's view hints of the consequential infliction that might be invoked upon the Borana, their livestock and other living and nonliving entities.

One might ask a logical question on how the Borana maintain and build peace within their community and between them, their neighbors and non-humans. The Borana maintain peace among themselves through the Gadaa system and the values and principles embedded within the system. Notions of peace, trust, reciprocity and coexistence are among the fundamental principle of the Gadaa system. For example, the *Buusaa-gonoffa* tradition, in which each clan member has a responsibility to support any member of the clan at times of natural and human-induced disasters, strengthens social capital, inter-group cohesion, sense of reciprocity and trust. According to Legesse (2000) and Adugna (2010), the Borana have strong harmony among themselves, despite recurrent conflict with their neighbors. Internally, the Borana have an intact cultural norm that prohibits aggression against their kin under any circumstance. In cases of dispute within clan members or between individuals, Gadaa leaders and local elders play a fundamental role in resolving conflicts and maintaining and building peace. In terms of inter-group relationship, principles of harmony, reciprocity and interdependence, which in turn build inter-trust, are core elements. However, this chapter does not romanticize traditional notions of inter-group relationship as such representation might overlook recurrent conflicts with their neighbors. Moreover, as

Borana elders succinctly explain, they are also not in complete harmony with nature. Drought, war, disease, intergenerational discrepancies in terms of respecting *Seera* Borana (Borana customary law) and other man-made and natural problems are attributed to and also signs of threatening the relationship between humans, nonhumans and supernatural power. In cases of conflict with their neighboring groups or in contexts of perceived change in harmony with nature, the Borana go through procedures of reconciliation that requires elders from Borana and their neighbors to undertake earnest rituals, dialogue and actions.

Epistemological and methodological considerations

In the preceding sections, I discussed that notions of peace, conflict and peacebuilding among the Borana are grounded in their ontological articulation of the place of humans in nature. I also discussed that the Borana conceptualize peace not as absence of conflict. Rather, it is a phenomenon, practice, action and knowledge that has multilayered dimensions and nurtured through different forms of social interaction, cultural values, dialogue and rituals. Like farmers who invest their knowledge, capital and technology from the period of land preparation to final harvest, the Borana also earnestly work on maintaining, nurturing and building peace. Issues of trust building, social capital, continuous rituals and dialogue are among key practices and actions the Borana consider important in the process of peacebuilding within the society and beyond.

In contrast to conflict research that is rooted in western epistemology of knowledge, indigenous epistemology subscribes to relational ways of knowing. Relational epistemology, as Chilisa (2012) argues, situates our spectrum of knowledge within relationships between humans, nonhumans and the cosmos, and treats all these dimensions as mutually constituted and culturally grounded.

Thus, any peacebuilding research in such socio-cultural settings should primarily take into consideration the ontological and epistemological positions of the society. More specifically, the conceptualization of peace and conflict as elements engrained within the tripartite relationship between humans, nonhumans (including the cosmos) and supernatural power hint upon the ontological disposition of the people – the question of humans' place in nature. Epistemologically too, peace and conflict are understood from humans' relations and interactions with other agents in their surrounding who, according to the Borana's relational epistemology for example, have the agency in shaping relationships. As conflict constitutes discursive phenomenon and actions, including absence of harmony between humans and nonhumans, peacebuilding also entails the continuous process of capitalizing on existing social capital, cultural values, mythological beliefs, principles enshrined in traditional institutions and culturally and economically grounded inter-group relations. Approaching the subject from relational perspective and situating notions and practices of peacebuilding within the cultural values, socio-economic contexts, spiritual and mythological orientations and traditional institutions of governance helps a nuanced analysis of changing dynamics, questions of sustainability and legitimacy of making and building peace.

Methodologically, researching peace and conflict poses different categories of methodological challenges. Broadly speaking, challenges to situate peace and conflict research within culturally embedded knowledge, issues of silence, ethnographic seduction and positionality make the subject more complex than other topics. As I discussed earlier in this chapter, indigenous societies' conceptualization of peace and conflict departs from the conventional Western notions, whereby, in the latter context, these concepts are understood merely as indicators of absence of violence and experiences of dispute/violence/disagreement, respectively. My fieldwork among the Borana reveals the need to develop robust methodological frameworks that address practices of peacebuilding, both from discursive and practical dimensions. It requires designing methodologies grounded within the epistemological positions of the people, their axiological perspectives – constituting values, customs, beliefs and worldviews. Concurrently, it also necessitates developing peace research methodologies that help us to disentangle competing perspectives, narratives, claims, contestations, actors, interests, institutions and actors' articulations of conflict and peacebuilding.

Conflict by its very nature brings different actors into contestation, and as a result, enhances dichotomies (Bank and Heur 2007). It also creates a sense of mistrust, animosity between parties/groups in conflict, while it also reduces the degree of confidence local communities have of researchers because there is often a tendency of associating researchers either to the "Other" or to the state. As a result, our research subjects prefer to be silent or resort to what James Scott (1990) states "public transcripts" preserving their genuine impressions, experiences and knowledge for "backstage" or "hidden transcripts". In contexts of political repression or fear of repression in speaking out about conflict, silence becomes a form of not only concealing information but also resistance to the system because of lack of trust. Likewise, the Borana were not initially open to talk about conflict, peacebuilding practices and other associated issues. Eventually, after long periods of rapport establishment and trust building, I managed to get backstage information on their conceptualization of inter-group conflict, peacebuilding process and their relations with the state.

As conflict creates dichotomies and polarizes inter-group relationships, research on themes like peace and conflict is subjected to our subjects' efforts to win the impression of researchers to their own side through a systematic strategy, what Roben (1995) calls "ethnographic seduction". Research subjects try to influence the course of the research process and its findings by asking researchers for compassion, collaboration and even complicity. Roben further argues that researchers in the field of peace and conflict studies should pragmatically approach the matter through *impression management*. One of the methodological challenges I encountered during my fieldwork among the Borana was related to my positionality as a researcher working on one's own community. Working on one's own ethnic group makes the researcher susceptible to challenges of positionality on the one hand, and pressure of ethnographic seduction on the other hand. My Borana informants were resolute to win my impression towards their suffering from recurrent aggression from the Somali who have been backed by successive Ethiopian regimes. As I clearly indicated in my earlier work, however, I used a diverse range of methods including subjects from conflict parties in Ethiopia and Kenya and also informants from groups not involved in the conflict to get impartial reflection on the subject (Regassa 2016). Methodological diversity and involving groups deemed neutral in the conflict help us reduce challenges of ethnographic seduction while impression management reduces researchers' subjectivity and positionality bias.

Conclusion

The Borana conceptualize peace as harmony between humans, nonhumans and the supernatural power, and it is nurtured, developed and maintained through values of trust, reciprocity and interdependence. The Borana maintain peace by investing in social capital, dialogue, institutional building and earnest prayer in which indigenous institutions such as the Gadaa system play a central role not only in creating social cohesion, but also setting values and principles in relationships among humans and between humans and nonhumans. As the Borana understand peace from a relational perspective, researching peacebuilding practices among the society should be grounded within the cultural settings and the people's ontological positions of their place in nature. While the presence of war or violence is one of the indications of absence of peace, the opposite does not necessarily guarantee peace in Borana land. Borana land is considered peaceful when the people live in harmony with their neighbors, their livestock are fertile, rain comes and goes timely, the youth respect the elderly and the people respect ancestral values and norms. Under such circumstances, researching peacebuilding should follow robust methodological paradigms grounded within relational epistemology that recognizes cultural, spiritual and mythological narratives, norms and values. Because conflict produces dichotomies between parties in conflict on the one hand, and parties on conflict try to win the impression of researchers towards their causes on the other hand, researching such topics (whether on conflict or peace building) requires developing refined methodological approaches.

Note

1 "Oromo: UNESCO recognizes Indigenous System of Self-Government as World Heritage," http://unpo.org/article/19701

References

Abbink, Jon. (1997) "Ethnicity and Constitutionalism in Contemporary Ethiopia," *Journal of African Law*, 41, 159–174.

Adugna, Fekadu. (2010) "Making Use of Kin Beyond the International Border: Inter-ethnic Relations along the Ethio-Kenyan Border," in Dereje Feyisa and Markus V. Hoehne (eds.), *Borders & Borderlands as Resources in the Horn of Africa*. New York: James Currey, 45–60.

Autesserre, Severine. (2010) *The Trouble with the Congo: Local Violence and the Failure of International Peace Building*. Cambridge: Cambridge University Press.

Bank, André and Heur, Bas Van. (2007) "Transnational Conflicts and the Politics of Scalar Networks: Evidence from Northern Africa," *Third World Quarterly*, 28 (3), 593–612.

Bassi, Marco. (2005). *Decisions in the Shade: Political and Juridical Processes among the Oromo Borana*. New York: Red Sea Press.

Bassi, Marco. (2010) "The Politics of Space in Borana Oromo, Ethiopia: Demographics, Elections, Identity and Customary Institutions," *Journal of Eastern African Studies*, 4 (2), 221–246.

Baxter, Paul. (1991) "Preface," in Van De Loo (ed.), *Guji Oromo Culture in Southern Ethiopia: Religious Capabilities in Rituals and Songs*. Berlin: Deitrich Reimer Verlag, 9–13.

Beriso, Taddesse. (2009) "Changing Alliances of Guji-Oromo and their Neighbors: State Policies and Local Factors," in Gunther Schlee and Elizabeth Wendy (eds.), *Identifications and Alliances in Northeast Africa*. London: Berghahn Books, 191–207.

Bhavnani, Ravi. (2009) "The Scarcity, Abundance, and Conflict: A Complex New World," *Whitehead Journal of Diplomacy and International Relations*, 19/20, 65–81.

Chilisa, Bagele. (2012) *Indigenous Research Methodologies*. Los Angeles, CA: Sage Publications.

Descola, Philip. (2013) *Beyond Nature and Culture*. Chicago, IL: The University of Chicago Press.

Hagmann, Tobias (2005) "Confronting the Concept of Environmentally Induced Conflict," *Peace, Conflict and Development: Issue Six*. https://www.files.ethz.ch/isn/15430/hagmann.pdf.

Homer-Dixon, Thomas. (1994) "Environmental Scarcities and Violent Conflict: Evidence from Cases," *International Security*, 19 (1), 5–40.

Hutchinson, Sharon E., and Pendle, Naomi R. (2015). "Violence, Legitimacy, and Prophecy: Nuer Struggles with Uncertainty in South Sudan," *American Ethnologist*, 42 (3), 415–430

Krätli, Saverio (2015). *Valuing Variability: New Perspectives on Climate Resilient Dryland Development*. London: International Institute for Environment and Development.

Legesse, Asmarom. (1973). *Gada: Three Approaches to the Study of African Society*. New York: Red Sea Press.

Legesse, Asmarom. (2000) *Oromo Democracy: An Indigenous African Political System*. Lawrenceville, NJ: Red Sea Press.

Omeje, Keneth and Hepner, Tricia. (2013) *Conflict and Peacebuilding in the African Great Lakes Region*. Bloomington: Indiana University Press.

Oba, Gufu. (2013) *Nomads in the Shadows of Empires*. Leiden and Boston, MA: Brill.

Regassa, Asebe. (2016) "Competing Orders and Conflicts on the Margins of the State: Inter-group Conflicts along Ethiopia-Kenya Border," *African Journal of Conflict Resolution* ACCORD, 16 (2), 57–84.

Regassa, A. Debelo, Legesse, Abiyot, Milstein, Tema, Orkaydo, Ongaye. (2017) "'Tree Is life': The Rising of Dualism and the Declining of Mutualism among the Gedeo of Southern Ethiopia," *Journal of Frontiers in Communication and Environmental Sciences*, https://doi.org/10.3389/fcomm.2017.00007.

Robben, A. (1995) "The Politics of Truth and Emotion among Victims and Perpetrators of Violence," in A. Roben and C. Nordstrom (eds.), *Fieldwork under Fire: Contemporary Studies of Violence and Survival*. Berkeley and London: University of California Press, pp. 81–104.

Schlee, Günther. (2003) "Redrawing the Map of the Horn: The Politics of Difference." *Journal of the International African Institute*, 73 (3), 343–368.

Scott, James. (1990) *Domination and the Arts of Resistance: Hidden Transcripts*. New Haven: Yale University Press.

Smith, Linda. (1999) *Decolonizing Methodologies: Research and Indigenous Peoples*. Dunedin: University of Otago Press.

Tache, Boku and Oba, Gufu. (2009) "Policy-driven Inter-ethnic Conflicts in Southern Ethiopia," *Review of African Political Economy* 36 (121), 409–426.

UNESCO. (2013) *The Role of Culture in Peace and Reconciliation*. Discussion Paper.

Zeleza, Paul Tiyambe. (2008) "Introduction: The Causes and Costs of War in Africa: From Liberation Struggles to the 'War on Terror'," in Alfred Nhema and Paul Tiyambe Zeleza (eds.), *The Roots of African Conflicts: The Causes and Costs*. Oxford: James Currey Ltd., pp. 1–5.

11 A critique of the conceptual documents that frame UN and AU practice of peacebuilding

Norman Sempijja

The United Nations Security Council's resolution S/PRST/2001/5 brought a new dimension to the definition of peacebuilding. It expands peacebuilding from a limited focus on explicit post-war activities to encompass activities and processes aimed at preventing the outbreak, the recurrence or continuation of armed conflict. At the same time, a variety of peace tools—conflict management, conflict resolution, peacebuilding and peace enforcement (Agenda for Peace)—have enabled hybrid approaches and fluid processes based on specific conflict contexts. Yet the proliferation and resurgence of conflicts have led to a need to rethink this approach to peacebuilding. This has necessitated for instance a need to accommodate experimental approaches infused by economic outputs rather than consolidating established peace cultures inclusive of, on the main, conflict management and resolution. With the largest number of peacekeeping operations being located in Africa, the continent is increasingly the site of experimentation for various mechanisms of intervention. This chapter examines and critiques three key documents that frame peacebuilding practice in Africa, namely An Agenda for Peace, the Brahimi report and the Capstone doctrine. It reflects in particular on AU peacebuilding practices with the aim of coming up with an alternative perspective to peacebuilding in Africa. Thus, it is the argument of this study that a new crucial dimension to peace processes is a plethora of experiments deployed by external agents who adopt a one-size-fit-all peacebuilding process as exemplified with the three documents being critiqued. This practice is often oblivious to factors like the nature of conflicts, processes of state formation and the inherent weaknesses within the neoliberal peacebuilding model leading to limited success in peacebuilding interventions.

Peacebuilding according to the UN is an evolving concept which first came to light when Johan Galtung called for the establishment of peacebuilding structures to promote sustainable peace by addressing the root causes of conflicts while 'supporting indigenous capacities for peace management and conflict resolution' (UN Peacebuilding Support Office, p. 5). The term peacebuilding was embraced by both Brahimi and Boutros-Boutros Ghali as they mapped a way forward in the post-Cold War era in an attempt to find ways to build sustainable peace. By and large, attempts at definition of peacebuilding have varied especially in terms of context. For example, Boutros-Boutros Ghali defined peacebuilding as a series of actions designed to solidify peace and avoid relapse into conflict (UN Peacebuilding Support Office, p. 5). This was quite general, but the key issues that stick out here are mainly related to stopping a relapse into conflict. But it does not give much detail on specific mechanisms of enforcement.

Brahimi expanded upon the definition above by defining peacebuilding as activities undertaken on the far side of conflict to reassemble the foundations of peace and provide tools for building on those foundations, something that is more than just the absence of war (UN Peacebuilding Support Office, p. 5). This definition buys into the arguments related to positive and negative peace as expounded by Johan Galtung, whereby negative peace refers to the absence of violence though it does not mean the absence of conflict (Galtung and Fischer 2013). So, Brahimi's definition builds on the positive peace argument, giving due attention to the legitimate needs and interests of all parties involved (Galtung and Fischer 2013). Other authors like Lederach (1997, p. 6) define peacebuilding as 'a multiplicity of interdependent elements and actions that contribute to the constructive transformation of the conflict.' His definition looks at peacebuilding as playing a transformative role in conflicts.

Cohen (2006) viewed peacebuilding in relation to stabilization and reconstruction, whereby the aim is to reduce violence and let key stake holders take control of the conflict resolution process and the rebuilding of the state or community involved. To Cohen, sources of violent conflict have to be 'reduced to such an extent that local institutions that are being developed have the capacity to resolve internal conflict peacefully and are able to prevail over violent extremists' (2006, p. 8). As for the Secretary-General's Policy Committee, it combines a number of perspectives (Transformative, reconstruction and development) in its definition of peacebuilding. The report argues that peacebuilding encompasses:

> a range of measures targeted to reduce the risk of lapsing or relapsing into conflict by strengthening national capacities at all levels for conflict management, and to lay the foundations for sustainable peace and development. Peacebuilding strategies must be coherent and tailored to the specific needs of the country concerned, based on national ownership, and should comprise a carefully prioritized, sequenced, and therefore relatively narrow set of activities aimed at achieving the above objectives.
> (UN Peacebuilding support office, pp. 5–6).

According to Campbell (2007, p. 4), successful peacebuilding can be viewed ideally in the context of absence of violent conflict and the existence of sustainable peace. Successful peacebuilding is normally both state-centered and people-centered. In his perspective, institutional reform should aim to address the specific grievances that fueled the conflict in the first place. On a people-centered level peacebuilding enables 'participants and communities to develop initiatives that decrease dividers increase connectors or address cusses of conflict' (ibid., p. 4).

The various debates that emerge from the definitions reflect a need, first of all, to end conflicts and prevent relapses. Second, the conception of a post-conflict society is based on institutional, market and democratic reforms which betray the authors' attachment to the liberal treatment of post-conflict societies. As Curtis (2013, pp. 85–86) and Newman et al. (2009, p. 4) note, post-conflict state building takes the form of a market state, which as will be shown below denies African states agency and is not on par with the needs of the states that are being reconstructed. This, according to Newman et al. (2009, p. 4), has led to questions being raised about whether a peacebuilding doctrine exists.

Thus far, the liberal peace thinking has been the dominant ideology in peacebuilding research and practice (Curtis 2013; Sabaratnam 2017). Although the centrality of a viable state has been acknowledged, there has been a struggle to reconcile this with the liberal market imperative that underpins the liberal peace framework. Curtis (2013) notes that even states like Liberia and Burundi that have been heralded as success stories are struggling both politically and economically. For instance, Burundi since 2015 has relapsed into low-intensity conflict after attempts by President Pierre Nkrunziza to stay in power beyond the agreed term (Council on Foreign Relations 2018).

Moreover, although the Democratic Republic Congo has held elections since 2006, this has not put an end to endemic conflicts and violent politics. The perspectives of Sabaratnam (2017) and Curtis (2013) are shared by Newman et al. (2009, p. 4) who contend that the efficacy and suitability of the application of liberal democracy and market economies in conflict-prone societies face serious scrutiny.

Therefore, when exploring the record of peacebuilding over the last 25 years, De Soto and Del Castillo (2016, p. 213) note that the record is bleak. More than half (12) of the 21 countries where the UN has launched peacebuilding missions have relapsed into conflict during the first 10 years of the transition. Yet, although some have avoided the relapse into conflict, the UN has had to maintain an expensive peacekeeping mission, as seen in the case of the Democratic Republic of Congo – the United Nations Organization Stabilization Mission in the Democratic Republic of Congo (MONUSCO), the largest and most expensive UN mission with personnel of 17,000 military troops, 1,500 police officers and 3,500 civilians with a budget of $1.14 billion for 2017–2018 (Coleman 2017).

Launched in 2003, the United Nations Mission in Liberia (UNMIL) was completed in March 2018. Since then, there has been a successful handover of power from President Sirleaf to Mr. George Weah in 2018. Lucey and Kumalo (2017, p. 1) note that the country still has peacebuilding priorities to be addressed. Furthermore, De Soto and del Castillo (2016, p. 214) argue that of the 21 countries where peacekeeping has taken place, the majority have been the worst performers, plus the peace index includes six of these countries as part of its worst eight performers. In addition, countries that have not relapsed into conflict or have had expensive peacekeeping missions are not doing well economically. For example, Mozambique is among the biggest recipients of aid and technical assistance (ibid., p. 214).

It is imperative therefore to examine the three critical documents that have been crucial to the development of peacebuilding to delve into what have been the key challenges and failures and to think about ways to improve it. As stated earlier, this chapter argues that a new crucial dimension to peace processes is a plethora of experiments deployed by external agents who adopt a one-size-fit-all peacebuilding process as exemplified with the three documents being critiqued.

This chapter builds on the work of authors Clapham (2002), Doner et al. (2005), Putzel (2005, p. 11) and Barbra (2008), who note that the peacebuilding model which has been adopted from liberal perspective has run into difficulties, especially as the application tends to be dogmatic and prematurely put into action plus there is an apparent lack of legitimacy. Besides, authors like Duffield (2007), Barnett (2011) and Charbonneau (2014) view liberal peacebuilding as imperial in nature and has an inherently paternalistic attitude of western humanitarianism. The study delves into the changing nature of peace mission in the post-Cold War; thereafter the Agenda for peace, Brahimi report and Capstone doctrine will be critiqued. This chapter will conclude with an exploration of possible ways to develop a more comprehensive and inclusive peacebuilding perspective.

Peacebuilding and the changing nature of peace missions in the post-Cold War era

Peacekeeping missions in the Cold War era were, according to Osmancavusoglu (1999–2000, p. 2), often entangled in local political dynamics. Osmancavusoglu argues that 'these forces became a long-term contributor to the civilian economy and provider of such services as medical treatment, housing, distribution of food and services, and family contacts' (ibid., p. 2). However, the post-Cold War era has witnessed a fresh wave of conflicts termed 'new wars' (Kaldor 2012, pp. 4–5) which have changed the role of peacekeeping. These wars, according to Melander et al. (2006, p. 7), are depicted by a failure to distinguish between internal and external, civilian and military categories and they are characteristically taking place in failed or failing states. These wars are mostly identity-based with a number of actors who do not exercise military restraint and have a disdain for ideological principles. In these wars, states are rarely able to subdue internal violence as the genocide in Rwanda attests (ibid., p. 10).

The phenomenon of new wars has been intensified by the proliferation of small arms made available in the aftermath of the Cold War. Kaldor (2012, p. 4) notes that these arms were originating from the successor states of the collapsed Soviet Union. Most of these arms found their way into the hands of both national armies and rebel factions in Third World countries. This development fanned conflict, as opposing sides were confident of victory and saw little need to resort to diplomatic solutions. Kaldor (2012, p. 5) argues further that the new wars are exacerbated by globalization. She contends that the global interconnectedness threatened the future of territory-based sovereignty, which is one of the major foundations of the modern state. The erosion of this sovereignty has undermined the autonomy of the state and in extreme circumstances has led to the disintegration of states.

This has arisen as a result of the erosion of states' monopoly over legitimate organized violence, primarily caused by the trans-nationalization of the military force. This practice began in World War II, was institutionalized by the bloc system in the Cold War and has transitioned into the post-Cold War period. Consequently, with increasing globalization, weak states have struggled to keep internal dissension under their control. Therefore, due to the aforementioned factors, peacekeeping has had to go through some drastic changes, especially as far as the role of peacekeepers is concerned.

Collective enforcement under the auspices of the UN was deeply challenged in the context of the Cold War. This led to the rise of observer missions (UNDPKO 2008, p. 1) and peacekeeping missions. Since they were not included in the original charter they came to be referred to as chapter VI and half as they fell 'somewhere between the traditional methods of resolving disputes peacefully (outlined in Chapter VI) on the one hand, and more forceful, less —consent-based action (Chapter VII), on the other' (ibid., p. 1).

With the demise of the Soviet Union leading to the end of the Cold War, the Security Council gained some leeway in the use of peacekeeping as a tool of conflict resolution in the international arena. But peacekeeping had to be refined in order to address the problems of the day because, as Szapiro (2004, pp. 348–349) notes, the 'end of the Cold War shifted the focus from strategic thinking from interstate ideologically fueled conflicts to predominantly intrastate multifaceted crises.' Szapiro (ibid., p. 349) further postulates that the lifting of the ideological veil shed a glaring light over the root causes of internal conflicts which involved a combination of degraded economic circumstances, social and political instability and weak states, among other factors. This therefore led to the development of a new conception (away from the Charter) on the meaning of security, from collective defense or territorial threat to human security based on the link between human rights, democracy, peace and development.

The earlier development consequently led to the revamping of the peacekeeping process as a whole, in that it became multifaceted. Apart from peacekeeping duties, a mission often incorporates peacebuilding in its agenda. For instance, peacekeeping missions are involved in overseeing the security sector reforms (SSR) of the recipient nation, carrying out elections, restructuring the judiciary and working with other humanitarian organizations to improve the economic, health and social situation of the recipient state. These processes were envisaged in the different UN documents, which will be explored below.

An agenda for peace

Contemporary peacebuilding research is linked to three important documents produced by the UN. These are 'An agenda for peace' and later 'supplement to an agenda for peace' by former Secretary-General Boutros-Boutros Ghali, the Brahimi Report and the Capstone Doctrine report. In *An Agenda for Peace*, Boutros Ghali admitted that the sources of conflicts in the post-Cold War era were pervasive and that serious efforts to pursue respect for human rights, fundamental freedoms and sustainable economic and social development were crucial to addressing them. He also argued that there was a need to alleviate distress and control the use of weapons of mass destruction (Ghali 1992, p. 1).

Most conflicts in the post-Cold War era were in fact intra-state in nature, and a large number were linked to contest on sharing state power, natural resources and ethnicity (Enuka 2012). Other drivers of conflicts were mainly attributed to social and economic problems in states leading to the rise of the security development nexus argument. The proponents of this argument state that lack of development foments insecurity. They also maintain that insecurity cannot create an adequate environment for development (Menkhaus 2004).

According to the above, security and development are not mutually exclusive. Therefore, to resolve conflicts, one has to factor in the social economic development of the population to bring about wider prosperity. The more people develop, the more likely they will be unwilling to engage in wars as they have something to protect. Hence, they are more willing to engage in diplomacy and non-violent means of agitation when required.

Chandler (2007) notes the argument by the German Federal Ministry of economic cooperation and development that 20 of the poorest countries in the world are caught up in conflict of an armed nature, which proves the point that poorer states are more prone to violent conflict than rich ones. Likewise, the World Bank and other institutions have shown that:

> a country with an annual per capita income of 250 US dollars has a 15 percent chance of civil war within five years, whereas in a country with an annual per capita income of 5,000 US dollars, the probability is less than 1 percent.
>
> (Chandler 2007, p. 363)

Boutros Ghali also advocated respect for human rights and fundamental freedoms; he thus promoted the development of institutions, constitutionalism, rule of law and regular elections. But he also stressed the need to protect religious, ethnic, social and linguistic minorities (Ghali 1992, p. 4). Consequently, he argued the UN had to assist peacebuilding in different contexts, for example:

> rebuilding the institutions and infrastructures of nations torn by civil war and strife; and building bonds of peaceful mutual benefit among nations formerly at war; And in the largest sense, to address the deepest causes of conflict: economic despair, social injustice and political oppression. It is possible to discern an increasingly common moral perception that spans the world's nations and peoples, and which is finding expression in international laws, many owing their genesis to the work of this Organization.
>
> (ibid., p. 4)

Thus, there is a need to strengthen new democratic institutions, as he believed that social peace is as important and strategic as political peace. Hence there is clear connection between 'democratic practices like rule of law and transparency in decision making and achievement of true peace and security in any new and stable political order' (ibid., p. 16).

Additionally, as the world was globalizing and becoming more integrated, Ghali noticed that the risks to stability in the international system were great. Apart from conflicts, the world faced dangers of unchecked population growth which would likely lead to more poverty and insecurity. This, in turn, would result into refugees, internally displaced people, oppression and barriers to trade, among other things (ibid., p. 3). Accordingly, post-conflict peacebuilding to him was defined as 'action to identify and support structures, which will tend to strengthen and solidify peace in order to avoid a relapse into conflict' (ibid., p. 5).

Yet even with the perceived merits of the peacebuilding model in the *Agenda for peace*, the UN has struggled immensely with rebuilding societies after conflict. In some places conflicts have re-emerged, as shown by Brosig and Sempijja (2017). For example, in Sierra Leone conflict kept reoccurring even with the deployment of peacekeepers and peacebuilding initiatives. In the Democratic Republic of Congo conflict persists even after 20 years of UN deployment. In other places, the UN has ended up staying longer, as the case is with Liberia and Democratic Republic of Congo, among others.

The other weakness of the peacebuilding model is highlighted by Call and Collin (2015) who argue that peacebuilding missions are often only launched at the end of the conflict as states are less willing to let interveners get involved in conflict-prevention strategies. This undermines the credibility of the process in the long run. From an operational perspective Súilleabháin (2015) and Call and Collin (2015) note that UN normally engages elites and is less inclined to use a bottom-up approach (involving local actors) thereby decreasing national ownership.

Thus the standard peacebuilding model has been seen as state-centric and has quite often sidelined non-state actors. Moreover it should be noted that most of the conflicts, for example, in Africa, are engineered by non-state actors. So keeping them out of the peacebuilding process quite often leads to failure and a resurgence of the conflict. The failure to use local knowledge robs the peacebuilding process of vital information that can lead to sustainable peace.

Furthermore, Barbara (2008, p. 310) noted that international efforts to construct viable states are undermined when the imperative of the neoliberal state building reform is applied in a premature fashion. For example, with a country in the process of economic recovery, there is a need for some protectionist measures to stabilize local industries. Nonetheless, these do not stand a chance with the arrival of multinational corporations. State-owned enterprises are crucial to supporting farmers; yet once removed many of them are left without subsidies to handle financial loss emanating from poor sales and natural disasters like famine.

Clapham (2002, pp. 778–779) builds on the above to note that there is a need for a solid material foundation when rebuilding states. This provides support for a 'Weberian infrastructure capable of maintaining order and building legitimacy' (ibid., pp. 778–779). This material support is crucial for states coming out of conflict to build legitimacy and work towards long-term independence. When the private sector is prematurely fostered, this can lead to failure and leave nascent states exposed.

Hence the building of legitimacy and long-term independence is crucial to forging a new social contract, especially through public employment and valued public goods and services. This thus helps respond to national political priorities (Doner et al. 2005). Moreover, Putzel (2005, p. 11) notes that the absence of private sector-led growth in newly 'rebuilt' neoliberal states leads to a lack of means 'to redress regional disparities or to promote redistributive policies that might reduce horizontal inequities' (Putzel 2005, p. 11).

Barbara (2008, p. 311) also noted that the largely dogmatic nature of the contemporary neoliberal development approaches further weakens the legitimacy of reformed political institutions such as the judiciary, the executive and the legislature. She contends that although state builders have the aforementioned institutions as the talisman of progress attained, these are hamstrung by the imposition of neoliberal policies as conditions for example of financial assistance. This denies institutions economic sovereignty and in the long run, political legitimacy as highlighted earlier.

Brahimi report: report of the panel on United Nations peace operations (2000)

In 1998, Lakhdar Brahimi was commissioned to head a special committee on peacekeeping and to explore the possibility of establishing post-conflict peacebuilding structures which would become part of the UN's arsenal of conflict resolution. The 2000 committee stressed the need to define and identify peacebuilding elements before being incorporated into the mandates of peace operations (Brahimi Report, p. 6)

The report called for the establishment of Peacebuilding to support UN peace operations. In some contexts, there can be independent initiatives with the express aim of supporting the consolidation of peace in post-conflict countries. The aim is to work with all stakeholders (Government and non-government parties). To the committee effective peacebuilding required 'active engagement with the local parties, and that engagement should be multidimensional in nature' (ibid., p. 6). However, as will be shown later, the call to engage with local parties has been without effects.

In line with An Agenda for Peace, the report called for free and fair elections in order to restore confidence in the ballot over the bullet (ibid., p. 6). But as Leininger (2006, p. 465) noted from a participant in a Lesson Learned conference in Haiti; 'You cannot have a one-year solution to a ten-year problem.' Leininger also recalled the saying among the Haitians that 'a constitution is paper, a bayonet is iron' (ibid., p. 465). This shows a sort of realism among post-conflict societies that merely having a constitution or holding elections does not spell the end of conflict.

Thus, elections previously seen as a hallmark of successful peacebuilding have proved a false dawn in countries like Angola, the Democratic Republic of Congo, South Sudan, Burundi and Afghanistan, among others. The winner-takes-all logic has been detrimental to former enemies as was in the Democratic Republic of Congo, when Kabila pushed out Bemba post-2006. Winners instrumentalize weak institutions for political and private ends. Hoove and Scholtbach (2008) argue that 'without strong political and state institutions, components of democratization such as elections and political party competition can raise tensions or, in extreme cases, lead to renewed conflict' (Hoove and Scholtbach 2008, p. 6).

The report called for UN civilian police to move away from merely documenting 'to reform[ing], train[ing] and restructure[ing] local police forces according to international standards for democratic policing and human rights, as well as having the capacity to respond effectively to civil disorder and for self-defence' (Brahimi Report, p. 6). Although Disarmament, Demobilization, Rehabilitation and Reintegration (DDRR) fulfilled the aim of removing weapons from the conflict and then demobilizing the soldiers in some cases and reintegrating them in society or retraining them to constitute a new police force or army, the process has had challenges in designing, sequencing and timing, among others. Indeed, Ball and Goor (2006) argue that DDRR processes:

> have the capacity to influence only a fairly narrow range of political and security objectives. They cannot substitute for inadequate will on the part of the parties to the conflict or peace enforcement activities. Nor can they prevent conflict from recurring. DDR also cannot produce development, guarantee the successful reintegration of ex-combatants into

> society, or substitute for longer-term programs to combat the proliferation of small arms and light weapons.
>
> (2006, p. 5)

Apart from the earlier point, the report recommended intervention by international judicial experts and human rights specialists in criminal system to help strengthen the rule of law despite the fact that prosecution is normally the reserve of a state. Consequently, a new government normally would go on with the recommendations of the UN, but in the long run would revert to its own practices. Likewise, traditional forms of justice have proved to be quite effective. For example where the international court of justice took years to prosecute a few individuals for war crimes, the Rwandan Gacaca courts (Clark and Kaufman 2009; Clark 2010) using traditional justice were able to clear the backlog of suspects who are believed to have participated in the genocide in a short time.

In a bid to form an effective forum for the formulation of peacebuilding strategies through the Executive Committee on Peace and Security (ECPS) the report invited institutions like:

> DPA, the Department of Political Affairs, the Department of Peacekeeping Operations (DPKO), the Office for the Coordination of Humanitarian Affairs, the Department of Disarmament Affairs, the Office of Legal Affairs, UNDP, the United Nations Children's Fund, OHCHR, UNHCR, the Special Representative of the Secretary- General for Children and Armed Conflict, and the United Nations Security Coordinator…the World Bank Group.

The Brahimi report was an improvement on the Agenda for peace in regard to the operationalizing of peacebuilding initiatives. The idea was to buttress the liberal peace perspective as it was seen as the way forward in resolving conflicts in the post-Cold War world. Yet the report fails to incorporate the key recipients of the peacebuilding initiatives, namely African states and regional bodies, which are the biggest targets of peacebuilding and relies instead on UN bodies and the World Bank.

Additionally, critics of the neoliberal peacebuilding model note a link between neoliberal peacebuilding model and imperialism. For example Charbonneau (2014, p. 611) drawing from Duffield (2007) explores the notions of *imperial peace* and the *liberal peace*, showing how both deployed a depoliticising humanitarian discourse of human development to impose, and/or police, order. The two are distinguished according to their purported goals: the *imperial peace* aspired to direct control while the *liberal peace* is said to aim at stability through non-territorial and networked relations of governance (Duffield 2007).

Charbonneau (2014, p. 610) notes that Barnett (2011) argues that given the historical roots and the inherent paternalistic attitude of Western humanitarianism, comparisons with imperialism are understandable, even predictable, for instance, French as the *gendarmerie* of Africa intervening in Cote d'Ivoire 2011, Libya 2011 and Mali 2013. France also used the EU as a cover in the Democratic Republic of Congo 2003, 2006, EUSEC and EUPOL and Chad/ Central African Republic 2008.

De Soto and Del Castillo (2016) noted that reducing the role of the state in the development agenda disempowers the people. Most of developed states like the United Kingdom, France and the United States employed a lot of state control in the beginning to protect internal market and industries. Even today European local farmers are subsidized heavily which makes locally produced products cheap. Hence the states being rebuilt after conflicts need to be able to benefit from such measures; otherwise development becomes superficial. For instance, the Democratic Republic of Congo is one of the fastest growing economies (Kuwonu 2016), but is among the bottom placed states on the human development index. For example, it was ranked among the lowest human development category 176 out of 186 in 2017 (United Nations Development Programme 2018).

Likewise, there is a disconnect between conflict resolution processes and actual peacebuilding; this has led to chaotic SSR within the police and army. For example, when exploring the SSR and human development in the police and army, the Democratic Republic of Congo, Brosig and Sempijja (2018) found that the structuring of the army and police post-conflict was not done in an orderly manner. Former combatants were given roles in either the police or the army to secure peace and also keep them employed in the post-conflict society.

On top of that, a degree of disorder arises from the fact that many actors with divergent interests are involved in the peacebuilding process. For example, the executive arm of government in a state being rebuilt may play the different organizations against each other to undermine other institutions. A case in point, Melmot (2009, p. 16) noted that competition was rife between the EU and the UN in the Democratic Republic of Congo mainly because both emulate each other to take the lead in the SSR.

On top of that, within certain organizations different states pursue different interests which may go against official policy or plan of action; these sometimes are referred to as lone riders like China, France and the United States seeking influence in Kinshasa to carry out military collaboration remotely from the reference structure and in an obscure manner (ibid., p. 16).

The Capstone doctrine

Although the Capstone document acknowledges that multidimensional UN peacekeeping operations may help to stem violence in the short term, it also points out that these do not necessarily lead to sustainable peace. It thus recommends accompanying peacekeeping with programs that prevent a recurrence of conflict. While there was an acknowledgement that there are conflicts that present unique challenges, it was argued that:

'experience has shown that the achievement of a sustainable peace requires progress in at least four critical areas:

a Restoring the State's ability to provide security and maintain public order;
b Strengthening the rule of law and respect for human rights;
c Supporting the emergence of legitimate political institutions and participatory processes;
d Promoting social and economic recovery and development, including the safe return or resettlement of internally displaced persons and refugees uprooted by conflict' (United Nations 2008, p. 25).

From the above, it is evident that the UN persists with a liberal interventionist approach[1] to peacekeeping, despite its obvious weaknesses as a framework for conflict resolution and sustainable peace.

The document highlights a desire to draw in other agencies to support DDR, mine action and reiterates the fact that successful implementation of SSR and Rule of law related activities is critical to defining the exit strategy of the UN. The latter thus aims to empower national security actors and institutions in order to help them function more effectively.

However, the one component that is ignored is the ability for the country to relapse into violence 10–15 years down the road. For example, Burundi typifies this retrogression whereby a leader decides to abrogate the constitution to achieve his ends.

Besides, there is no clear yardstick as to what constitutes functioning security institutions. But effectiveness seems to be understood in terms of an ability to protect the state from negative forces. If this is the case, then further queries will arise as to whether institutions are equipped to resist manipulation by a leader; hence the question of the independence of the institutions.

The other questions center on institutional reconstruction. For example, in some countries like Liberia, an education criterion was set in motion for recruiting soldiers and the police force. In contrast, reconstituted armed forces in the Democratic Republic of Congo were an amalgamation of former combatants, and this posed problems for President Joseph Kabila who was faced with an increasingly fragmented army. He could not rely on an army that was partly controlled by his former enemies (Barnet 2016). Promotion and demotion come at the whim of the president. There is, for instance, no separation of the chain of payment from the chain of command in the Democratic Republic of Congo army; soldiers are not paid for months and end up engaging in corruption in order to get by (Brosig and Sempijja 2018). Accordingly, rebuilding the security institutions is indeed a key requirement of peacebuilding, but liaison with country-specific practices on composition and managing of institutions needs to be factored in.

On gender-based violence in the army and police, the patriarchal system prevalent in most societies in Africa has made it difficult to apply principles of fair treatment of women in armed forces and the police despite noted UN efforts to eradicate gender-based discrimination (Ramisetty and Muriu 2013; Brosig and Sempijja 2018).

In regards to restoring state authority over territorial control (United Nations 2008, p. 27), Brosig and Sempijja (2017) note that UN peacekeeping missions have improved the national security of targeted states. Still, the UN has faced challenges in developing political participation, especially through electoral processes. Most elections that have followed UN peacebuilding have not been free and fair and they have not necessarily led to a resolution of conflicts. As noted in the Brahimi report, elections do not signify an end to conflict. At times, they have merely legitimized incumbents to obliterate their opposition, as seen in Burundi 2015 and the Democratic Republic of Congo post-2006 and 2011.

Thus, although Fath-Lihic and Brancati (2017, pp. 1, 3) argue that 'elections are key for fostering peace and stability in transitional settings and are often, as a result, one of the central elements of peace deals and transitional political accords,' they rightly point out that timing and sequencing of transitional elections are key to successful transition from conflict to sustainable peace.

Towards a more comprehensive and inclusive perspective on peacebuilding

There has been widespread acceptance of the inadequacies of the neoliberal model of peacebuilding; for example, Barbara (2008) and Charbonneau (2014), among others, have highlighted the key weaknesses of this approach as explored above. In keeping with the discreditation of the neoliberal policies, Anam (2018) and Richmond (2009) have proposed a hybrid peace approach which incorporates local dynamics. However, although these authors highlight the need to change the focus of the peacebuilding, they maintain the neoliberal outlook in peacebuilding.

Forsyth et al. (2017, p. 3), in their exploration of hybrid peacebuilding, note that it seeks to capture the 'intertwined relationship between the global and the local, the formal and the informal and the liberal and the illiberal.' Therefore, this characterizes what ought to be the actual practice of contemporary peacebuilding as opposed to the assumptions of the underpinning (liberal peace) theory. It is argued that the outcome of these interactions – the 'hybrid peace' – is both a more accurate depiction of the complex realities on the ground and more legitimate than the liberal peace because it taps into local agency and knowledge, thereby broadening the peace constituency and more effectively capturing the dynamic and interactive processes involved.

In highlighting the challenges of the neoliberal peacebuilding paradigm, Richmond (2009, p. 326) argues that neoliberal peace has led to the 'hybridized form of liberal/authoritarian state; this is symptomatic of an uncomfortable confrontation and relationship between the agents of the liberal peace and often, the agents of the conflict itself.' This has therefore led to a liberal anxiety directed both inwards and outwards. In the long run the needs and lives of the common person are obfuscated; yet one of the main aims of liberalism is to improve the well-being of people to give the institutional framework an opportunity to operate well, leading to the kind of peace dividend discussed earlier (Richmond (2009, p. 326).

Accordingly, Richmond further reinforces his perspective by arguing that peace is something organic: it rests on hybridity, the everyday, agency and self-determination. To him, organic peace represents politics before ideology, which implies the prominence of needs and everyday contexts for the common people (pp. 328–329; Richmond 2009).

Lisa Philipsen (2014) moves the debate further by arguing for integrated peacebuilding. This, she notes, is based on three pillars, namely national ownership, accountability and integration. The first pillar of ownership she alludes to is the agenda for change in Sierra Leone. This is in line with proposals by authors like Súilleabháin (2015, p. iii), who argue for an increased role for local actors in peacebuilding. She highlights the need for local actors and local ownership during peacebuilding, the need for local knowledge which is normally used as a passive source of input. She also argues for a need to map peace resources and networks in local communities.

Likewise, when it comes to selecting local partners, the international community engages only the elites in the cities. According to Súilleabháin:

> [t]his inhibits deeper buy-in for projects implemented in communities without consultation on priorities and program design. Still, many civil society groups in Africa face challenges of internal governance, representativeness, and legitimacy. As the case studies demonstrate, community-level and grassroots actors do not speak with one voice, and they are not all or always committed to peace.
>
> (2015, pp. iii–iv)

Súilleabháin asserts that the higher the levels of inclusivity in peacebuilding, the more sustainable peace will be achieved. Hence, inclusion of civil societies and local groups is fundamental to lasting peace.

The accountability pillar, according to Philipsen (2014), is centered around building strong state and accountable institutions of government. While integration should involve state and non-state actors in the peacebuilding process, Philipsen and Richmond, among others, would like to see the neoliberal trajectory of peacebuilding preserved. However, it is the proposition of the study for an integrated peacebuilding paradigm that provides a stronger role for regional bodies and at the same time pays close attention to the decolonization perspective. This is important, Zondi (2017) notes, as a need to reconstruct states minus the colonial structures.

Thus, the new integrated peacebuilding paradigm builds on Philipsen's (2014) perspective with stronger emphasis, for example, for more role of the AU and a resetting of the button for post-conflict societies to rebuild societies on bespoke ideals. This does not mean that a neoliberal approach should be abandoned; however, the people should have agency in deciding on how they will apply the neoliberal structures. As shown by Richmond, where peace was realized 'organically', the building of post-conflict societies should be done organically.

The hybrid peacebuilding, as proposed by Richmond, still does not give agency to the people in post-conflict societies. Instead, the problem is diagnosed for the people and the solutions proposed for them. Yet it should be the people that diagnose their problem and design solutions. In that way, they have ownership of the process. It should be noted that African states never got rid of the colonial state. Most states merely retained the oppressive apparatus of the colonial state and clothed it in new country name, constitution and leadership. Yet colonial laws were retained and some are being used in the penal codes today. Most conflicts in Africa are related to the contestation of the post-colonial state governance and questions like who should be a citizen, who should govern. Therefore, the imposition of the neoliberal apparatus on post-conflict societies, where such questions linger only postpones the recourse to conflict, also creates new problems.

Conclusion

In conclusion, therefore, the conceptualization and application of peacebuilding should not just be an external endeavor but should involve stakeholders. Although the neoliberal peacebuilding model has come under criticism, different scholars have come up with ways of accommodating its principles in conceiving the reconstruction of post-conflict societies, as seen in the case of Richmond and Philipsen. A non-dogmatic application of the concept that caters for agency could provide viable and long-lasting results. Nevertheless, the focus needs to move towards building sustainable peace, especially from a human development and security perspective. In the specific case of Africa, further consideration should be made concerning the decolonization process. As explored earlier, most contestation of the African state is deeply rooted in the way in which the post-colonial state was structured. Moreover, there is a need for international bodies to move away from working in silos and instead develop models of peacebuilding implementation that bring all actors together to achieve common goals that are of benefit to recipients of interventions. Thus, although the restructuring of the state around liberal norms and embracing of democratic principles is not entirely unproblematic, it is important to develop bespoke models of state building as all states are not structured in the same way and respond differently to different policies.

References

Anam, S. (2018). Peacebuilding: The Shift towards a Hybrid Peace Approach. *Global Strategies*, *9*(1), 37–48.

Ball, N. and Van de Goor, L. (2006). Disarmament, Demobilization and Reintegration: Mapping Issues, Dilemmas and Guiding Principles. Netherlands Institute of International Relations 'Clingendael', Conflict Research Unit. Available at http://www. clingendael.nl/publications/2006/20060800_cru_paper_ddr. pdf.

Barbara, J. (2008). Rethinking Neo-liberal State Building: Building Post-conflict Development States. *Development in Practice*, *18*(3), 307–318.

Barnet, James (2016). DR Congo in Crisis: Can Kabila Trust His Own Army? African Arguments. Available at http://africanarguments.org/2016/09/20/dr-congo-in-crisis-can-kabila-trust-his-own-army/ (accessed 12/12/2017).

Barnett, M. (2011). *Empire of Humanity: A History of Humanitarianism*. New York: Cornell University Press.

Boutros-Ghali, B. (1992). An Agenda For Peace: Preventive Diplomacy, Peacemaking and Peace-keeping. *International Relations*, *11*(3), 201–218.

Brosig, M., and Sempijja, N. (July 2015). The African Capacity for Immediate Response to Crisis: Workshop Report and Policy Advice. *South African institute of International Affairs Policy Insights*, *22*, 1–8.

Brosig, M., and Sempijja, N. (2017). What Peacekeeping Leaves Behind: Evaluating the Effects of Multi-dimensional Peace Operations in Africa. *Conflict, Security & Development*, *17*(1), 21–52.

Brosig, M., and Sempijja, N. (2018). Human Development and Security Sector Reform: The Examples of Liberia and the Democratic Republic of Congo. *African Security*, *11*(1), 56–83.

Call, C. T. and Collin, K. (2015). The United Nations Approach to Peacebuilding. Paper Prepared for 'Rising Powers and Peacebuilding' Project.

Campbell, S. (2007). What Is Successful Peacebuilding? A Report Prepared for Catholic Relief Services. Available at http://www.susannacampbell.com/wp-content/uploads/2016/02/Campbell-FINAL-CRS-Report-2007.pdf (accessed 1/1/2018).

Chandler, D. (2007). The Security–development Nexus and the Rise of 'anti-foreign policy'. *Journal of International Relations and Development*, *10*(4), 362–386.

Charbonneau, B. (2014). The Imperial Legacy of International Peacebuilding: The Case of Francophone Africa. *Review of International Studies*, *40*(3), 607–630.

Clapham, C. (2002). The Challenge to the State in a Globalized World. *Development and Change*, *33*(5), 775–795.

Clark, Phil. (2010). *The Gacaca Courts, Post-genocide Justice and Reconciliation in Rwanda: Justice without Lawyers*. Cambridge: Cambridge University Press.

Clark, P. and Kaufman, Z. K. (2009). *After Genocide: Transitional Justice, Post-conflict Reconstruction and Reconciliation in Rwanda and Beyond*. New York: Columbia University Press.

Cohen, C. (2006). Measuring Progress in Stabilization and Reconstruction (No. 1). United States Institute of Peace. United States Institute of Peace Special Report, Stabilisation and Reconstruction series No. 1.

Coleman, K. P. (2017). The Dynamics of Peacekeeping Budget Cuts: The Case of MONUSCO. *IPI Global Observatory*. Available at https://theglobalobservatory.org/by/katharina-p-coleman/ (accessed 27/12/2017).

Council on Foreign Relations. (2018). Political Crisis in Burundi. Global Conflict Tracker. Available at https://www.cfr.org/interactives/global-conflict-tracker?marker=23#!/conflict/political-crisis-in-burundi (accessed 4/6/2018).

Curtis, D. (2013). The Limits to Statebuilding for Peace in Africa. *South African Journal of International Affairs, 20*(1), 79–97.

De Soto, A. and Del Castillo, G. (2016). Obstacles to Peacebuilding Revisited. Global Governance. *A Review of Multilateralism and International Organizations*, April–June 2016, *22*(2), 209–227.

Doner, R. F., Ritchie, B. K., and Slater, D. (2005). Systemic Vulnerability and the Origins of Developmental States: Northeast and Southeast Asia in Comparative Perspective. *International Organization, 59*(2), 327–361.

Duffield, M. (2007). *Development, Security and Unending War: Governing the World of Peoples.* Cambridge: Polity Press.

Enuka, Chuka. (2012). Post-Cold War Conflicts: Imperative for Armed Humanitarian Intervention. *Global Journal of Human Social Sciences Interdisciplinary, 12*(9), 17–26.

Fath-Lihic and Brancati. (2017). Elections and Peacebuilding: Why the Timing and Sequencing of Transitional Elections Matter. Kofi Anan Foundation Policy Brief n.4. Available at http://www.kofiannanfoundation.org/app/uploads/2017/05/Elections-and-Peacebuilding.pdf (accessed (11/12/2017).

Forsyth, M., Kent, L., Dinnen, S., Wallis, J., and Bose, S. (2017). Hybridity in Peacebuilding and Development: A Critical Approach. *Third World Thematics: A TWQ Journal, 2*(4), 407–421.

Galtung J., and Fischer D. (2013) Positive and Negative Peace. In Johan Galtung (eds.), *SpringerBriefs on Pioneers in Science and Practice*, volume 5, pp. 173–178. Springer, Berlin, Heidelberg.

Kaldor, M. (2012). *New and Old Wars. Organized Violence in a Global Era.* Cambridge: Polity Press.

Kuwonu, F. (2016). DRC Economy: The Giant Awakens. *Africa Renewal, 30*(2), 22–23.

Lederach, J.P. (1997). *Sustainable Reconciliation in Divided Societies.* Washington, DC: USIP.

Leininger, J. (2006). Democracy and UN Peace-keeping-Conflict Resolution through State-building and Democracy Promotion in Haiti. In A. von Bogdandy and R. Wolfrum (eds.), *Max Planck Yearbook of United Nations Law*, Volume 10, pp. 465–530. Koninklijke: Brill N.V. Printed in The Netherlands.

Lotte, ten Hoove, Scholtbach, A. P., and Maul-Phillips, J. (2008). *Democracy and Political Party Assistance in Post-Conflict Societies.* The Hague: Netherland Institute of Multiparty Democracy Knowledge Centre.

Lucey, A. and Kumalo, L. (2017). Partnering for Sustainable Peace in Liberia. *Institute for Security Studies Papers, 2017*(302), 1–20.

Melander, E., Öberg, M., and Hall, J. (2006). *The "new Wars" Debate Revisited: An Empirical Evaluation of the Atrociousness of "new Wars".* Uppsala: Uppsala University, Department of Peace and Conflict Research.

Melmot, S. (2009). Candide au Congo L'échec annoncé de la réforme du secteur de sécurité (RSS). Focus stratégique n°9, Laboratoire de Recherche sur la Défense. Paris: IFRI.

Menkhaus, K. (2004). Vicious Circles and the Security Development Nexus in Somalia. *Conflict, Security & Development, 4*(2), 149–165.

Newman, E., Paris, R., and Richmond, O. P. (Eds.). (2009). *New Perspectives on Liberal Peacebuilding*. Tokyo: United Nations University Press.

Osmancavusoglu, E. (1999). Challenges to United Nations Peacekeeping Operations in the Post-Cold War Era. *Perceptions Journal of International Affairs, 4*(4), 1–8.

Philipsen, L. (2014). When Liberal Peacebuilding Fails: Paradoxes of Implementing Ownership and Accountability in the Integrated Approach. *Journal of Intervention and Statebuilding, 8*(1), 42–67.

Putzel, J. (2005). Globalization, Liberalization, and Prospects for the State. *International Political Science Review, 26*(1), 5–16.

Ramisetty, A. and Muriu, M. (2013). When Does the End Begin? Addressing Gender-based Violence in Post-conflict Societies: Case Studies from Zimbabwe and El Salvador. *Gender & Development, 21*(3), 489–503.

Richmond, O. P. (2009). Becoming Liberal, Unbecoming Liberalism: Liberal-local Hybridity Via the Everyday as a Response to the Paradoxes of Liberal Peacebuilding. *Journal of Intervention and Statebuilding, 3*(3), 324–344.

Sabaratnam, M. (2017). *Decolonising Intervention: International Statebuilding in Mozambique*. London: Rowman & Littlefield International.

Szapiro, M. (2004). International Organisations and Cooperation in Field of Conflict Prevention. In Vincent Kronenberger and Jan Wouters (eds.), *The EU and Conflict Prevention Policy and Legal Aspect*, pp. 347–368. The Hague: TMC Asser Press.

Súilleabháin, A. Ó. (Ed.). (2015). *Leveraging Local Knowledge for Peacebuilding and Statebuilding in Africa*. New York: International Peace Institute.

United Nations. (2008). United Nations Peacekeeping Operations Principles and Guidelines (Capstone). New York. Available at https://peacekeeping.un.org/sites/default/files/capstone_eng_0.pdf (accessed 10/11/2017).

United Nations Department of Peacekeeping. (2008). 60 Years of United Nations Peacekeeping, May 2008.

United Nations Development Programme (2018). Human Development Indices and Indicators: 2018 Statistical Update; Briefing note for countries on the 2018 Statistical Update' Democratic Republic of Congo. Available at http://hdr.undp.org/sites/default/files/Country-Profiles/COD.pdf (accessed 8/9/2019)

United Nations General Assembly and Security Council. (2000). Report of the Panel on United Nations Peace Operations (A/55/305 S/2000/809), 21 August 2000 (Brahimi Report). Available at http://www.un.org/documents/ga/docs/55/a55305.pdf (accessed 10/11/2017).

UN Peacebuilding Support Office; UN Peacebuilding: An Orientation. New York.

Zondi, S. (2017). African Union Approaches to Peacebuilding; Report from African Centre for the Constructive Resolution of Disputes. Available at https://reliefweb.int/report/world/african-union-approaches-peacebuilding (accessed 1/2/2018).

12 Doing research in the 'silicon savannah'

Digital media, peacebuilding and constructions of (in) stability in Kenya

Duncan Omanga and Pamela Chepngetich Mainye

In 2013, we were part of a team that won a research grant generously funded by a top North American Communications School. The call for proposals titled 'ICTs and Peacebuilding' that preceded the actual award had all the indications of an open, competitive process. We were to establish to what extent various digital technologies were deployed to map and monitor electoral-related violence in parts of Nairobi in the 2013 general elections, which saw the then International Criminal Court (ICC) indictees Uhuru Kenyatta and William Ruto elected into the presidency. The project drew from a unique context of contrarieties: Kenya had experienced a serious bout of election-related violence in the previous poll in 2007 and yet, in that period, had built a name as a hub of digital innovations in the region. Our aim was to trace the efficacies of these solutions to peacebuilding prior to and during the election week of 2013 and to trace specific end users of digital technologies. During the course of this project, we discovered a complex tapestry of power hierarchies, in which Africa is often constructed through studies that study 'down' to marginal and disempowered slam communities, and where a specific, deliberate construction of digital 'successes' from a network of local and external actors within the fledgling 'Nairobi tech' community flourished.

The grant providers asked us to play down these discoveries and instead focus on the 'end users' profiles and 'experiences'. The 'invisibility' of what we felt was 'genuine' end users to digital media and peacebuilding initiatives in Kenya led to a premature termination of the grant. Our conclusion was that the efficacy of peacebuilding initiatives through digital media technologies will only be fully realised if such initiatives are bottom-up, initiated from below rather than the prevailing digitisation within a 'foreign donor-aid logic'.

Introduction

Despite the adoption of a new constitution in 2010, the post-election violence surrounding the flawed 2007 General Elections have fuelled ongoing debates in Kenya about a state of political crisis and fragile democracy. Although Kenya has been experiencing repeated cycles of violence every election year since the inception of multi-party politics in the country, the 2007 violence reached levels that had never before been experienced. Over 1,000 people were killed and about 300,000 others displaced from their homes. As the violence unfolded, an interesting, if not innovative, idea was born out of this crisis by a Kenyan lady somehow caught up in the violence. At the height of the post-election violence of 2007/2008, *Ushahidi*, a digital media platform that was created to map incidents of violence occurring in the country as shared by citizens via the web and text was created. The name itself, Ushahidi (Swahili for testimony), illustrates the central role of citizens on the platform. Citizens were to give 'testimony', informing others on the happenings on the ground, alerting authorities and enabling faster response (Omenya and Crandall, 2013, p. 30).

Ushahidi has since been exported to different countries as a model for crisis response, human rights reporting and election monitoring. According to the Wikipedia page entry, *Ushahidi*:

> ...uses the concept of crowdsourcing for social activism and public accountability, serving as an initial model for what has been coined as 'activist mapping'—the combination of social activism, citizen journalism and geospatial information. *Ushahidi* offers products that enable local observers to submit reports using their mobile phones or the internet, while simultaneously creating a temporal and geospatial archive of events[1]

Prior to the 2013 elections, and with the ghosts of the 2007 elections hovering on Kenya, there was a cautiousness that constituted a general consensus from the state. Established bodies such as the church and the mainstream media concurred that the political discourse before and after the elections needed to be more carefully managed. Naturally, there was a curious disconnect between what formed the news headlines in the leading papers and media houses and actual campaign rhetoric. It was clear that the mainstream media embodied the phenomenon of inscribed 'silences' (Ugangu, 2011), which though understandable, might have empowered and greatly given space to online platforms such as the social media and other ICT initiatives.

Meanwhile, the silences from the mainstream media as a site of political discourse produced a new empty space that was filled by particular individuals and groups, who as ordinary citizens, human rights activists and techies became exemplars of a new aggressive, bold attempt at monitoring, reporting and sharing information on the elections and of calling the political class to account through diverse ICTs. Months before the 2013 elections, similar initiatives saw bloggers, self-proclaimed techies with substantial online 'influence' through significant online 'following' and crowdsourcing initiatives playing a crucial role in reporting and monitoring instances of elections-related violence. Among these, SMS sharing through mobile telephony, crowdsourcing platforms (both state-initiated and privately initiated platforms) and the social media provided sites through which diverse forms of experimentation with regard to Kenya's 2013 elections were done.

This chapter focuses largely on crowdsourcing digital platforms that were heavily funded by Western patrons, and domiciled in the two biggest slums in Nairobi—Kibera and Mathare. It reveals the complexities of doing research on digital technologies as it relates to peacebuilding in Africa, where the interests of the grant providers appear to overlap with those of 'techies' on the ground, to deliberately reproduce a discourse of digital success in developing countries. Using data from the field and monitoring of online activity during the course of the research, this research reveals how research on building in Africa is at risk of being 'social constructed' and 'dominated' to promote specific lines of thinking and policy framework within a liberal peacebuilding framework.

The rationale of this study comes in the backdrop of what appears to be fabulous claims of how digital media has changed the media ecology and social practice in countries in the south like Kenya and a characteristic 'Africa rising' narrative.[2] From our own experiences during this research, we argue that specific categories of actors in the whole dynamic of creating, circulating and utilising digital media in Africa have unique and often differing interests that determine the level and nature of participation in these platforms. Accordingly, the digital ecology, created by the confluence of digital media, personal interests, the dynamics of donor agenda and constructions of the African political space regarding violence, reproduce a symbolic space that mirrors these interests more than the actual impact of digital technologies on peacebuilding in Africa.

Postman (1970) was among the first media scholars to explicitly call for an effort to study media as environments, arguing that environments structure what we can see, say and do. They also assign roles and pressure us to play them. In like manner, media environments specify what we can do and what we cannot do (Scolari, 2012, p. 205). In the study of the media in Africa, several scholars have already made tentative investigations into the communicative ecology of media (Price and Stremlau, 2012; Tacchi et al., 2003). In these studies, there is a recurring pattern of taking a 'holistic' view of the interactions between media and their users. Generally, the media ecology envisaged in this study explores the kind of roles particular media force, or position individuals to play, and how the media structures ways of seeing and thinking, and even prompting particular actions (and thus seen, assumed or believed to be building peace). We also draw from critical theory to evaluate and illuminate the digital field occupied by 'digital patrons', slum communities and discourses of digital successes. This will not be the first study to question and probe the role of digital media to social issues, such as peacebuilding and conflict in Africa (Omanga and Omanga-Chepngetich, 2013; Omanga, 2015; Molony, 2018). Several other scholars have argued that there is a need to be cautious in making broad claims of the impact of ICTs and new media uses in Africa (Berger, 2002; Nyabuga, 2008; Mudhai et al., 2009; Asiedu, 2012).

Observations were drawn mainly from the deployment of 'digital solutions' to the 2013 elections, focusing on the three most prominent crowdsourcing sites in Kenya then: *Uchaguzi, Voice of Kibera* and *Voice of Mathare*. Further, we show how our 'near impossible' attempts at locating the end users of these platforms through interviews ended up mostly with civil groups working with slum communities (who harnessed and drove publicity efforts among potential users) and with administrators of blogs/crowdsourcing platforms than the actual, intended end users.

Digital media and the 2013 elections in Kenya

The near obsession of linking peace efforts around presidential elections to digital media technologies draws from a substantive historical and social context. According to the latest statistics from the Communications Authority of Kenya, mobile phone subscriptions stand at slightly over 38.9 million out of a population of about 43 million people (87 percent). Additionally, a similar number of Kenyans owning phones have some access or subscriptions to the internet (87.9 percent)[3]. These are extremely high figures for an African country. The gains made in the ICT sector in Kenya have been recognised globally over the past five years and Kenya's pioneering mobile innovation technologies have certainly propelled the country onto the global stage. According to a paper presented by Dr. Bitange Ndemo at St. Anthony's College and later published in the *Stability Journal* (Ndemo, 2015), this growth can perhaps be attributed to the path Kenya took in ICT development, in which it pursued an open access approach.

The majority of East African countries have some control over the Internet, preferring to pursue a centralised model that hinders innovation and creativity. In Kenya, open access to the Internet has revolutionised the way individuals communicate and collaborate, how entrepreneurs and corporations conduct business and how governments and citizens interact. At the same time, the Internet has established a revolutionary open model for its own development and governance, encompassing several stakeholders. In the past few years Kenya's creativity in mobile applications has gained much attention. Financial applications such as M-pesa as well as initiatives in agriculture, health and education are creating some remarkable strides, to the extent that researchers are asking why is it that these new applications are largely coming from Kenya and not any of its neighbours or the developed countries. It is in this context that Kenya has been described as the 'silicon Savannah'.

Ushahidi had already carved its place in 2007 as a platform that could monitor election-related violence. While the actual power of *Ushahidi* drew more from narratives of Afro-optimism and Africa rising in a context of unprecedented electoral violence, coupled with an influential online presence of Diasporic, 'Kenyan' Afropolitans, the actual impact on the ground was negligible, if not impossible to determine. The 'successes' of *Ushahidi* were and continue to be hailed as a unique African story, evidence that innovation can also flow from the south to the global north. The consistent mention of some of the 'founders' of the platform as among the most influential Africans in publications in the West has only served to anchor this construction[4]. During his visit to Kenya, US President Barack Obama highlighted in his speech Kenya's unique contribution to digital innovation in Africa and named the founders of *Ushahidi* as a beacon of African modernity. In such a context, the 2013 elections were possibly the most 'digitised' with regard to monitoring and mapping violence, especially in Nairobi.

Prior to the 2013 general elections, Kenya had witnessed a 'digital turn' in monitoring election-related violence, well dramatised through networks of funding, government policy and donor-civil society relations. The most readily recognisable of these platforms was *Uchaguzi* (Swahili for elections), a project following the path of the *Ushahidi* initiative (and initiated by the very same group behind *Ushahidi*). With a life span of just a week, it was designed to have a short but specific impact during the elections. It had already been piloted during Kenya's Constitutional referendum in 2010. As a platform, it engaged end users through SMS at the short code 3002 and also via social media *#uchaguzi* and via web forms. All these posts were designed to bounce off these codes and appear in a coherent format on the *Uchaguzi* webpage. According to the 'I-hub' blog (the creators of both *Ushahidi* and *Uchaguzi*), as a crowdsourcing platform:

> *Uchaguzi* is a technology platform that allows citizens and civil society to monitor and report incidences around the electoral process. *Uchaguzi* provides web and mobile-based channels for citizens and civil society to report on electoral offences such as intimidation, hate speech, vote buying, polling clerk bias, voting mis-information etc. The reports are then sent to the electoral authorities or security personnel for action.[5, 6]

There were other platforms that were specifically targeted to marginal communities in the slums and/or places considered to be politically volatile. There was *Voice of Kibera* and *Voice of Mathare*, both projects of the *Map of Kibera Trust*, aimed at monitoring only electoral events happening in the sprawling slums of Kibera and Mathare. The projects also had a special SMS number and also had a web form to post to. A similar initiative was undertaken by the *SiSi Ni Amani*-Kenya (SNA-K) (meaning, We are Peace, Kenya), an organisation which worked with a number of local peace groups to set up an SMS-based programming available to subscribers through a special code. Subscribers were able to dial in for free from any mobile phone line to subscribe and receive SMS from SNA-K. The project aimed at looking at rumours spreading via SMS and had a team of what they described as 'peace-keepers' on the ground responding timely to it by directly addressing the problem.

These initiatives were not limited to the private sector only, but were also taken up by various state agencies. For instance, in the Ministry of State, a platform known as the *Amani Kenya* @108 was launched (Amani means peace in Swahili). This was a project from a Committee on peacebuilding and conflict management under the Ministry of state for internal security. The system made use of setups known as District Peace Committees (DPCs), Peace Monitors and other relevant parties to gather crucial information from the field. Once information was gathered from various sources on the field, an analysis group was able to analyse the information and to issue an indicator-based Early Warning Report to the relevant parties for a response. Amani had its own short code for reporting on election-related events. In addition to these there was also a web form to report to online. Similarly, the country's electoral body, The Independent Electoral and Boundaries Commission (IEBC) had a Whistle Blowing Portal, where people reported via web any issue of competency to the Director Risk and Compliance, IEBC. They also had election hotlines for issues, complaints or inquiries.

Figure 12.1 shows the front interface of the *Voice of Kibera* platform, which allows users (residents) to post reports (known as citizens' reports) either virtually or using their phones. Any text sent to the mobile phone provided on the right of the image is then translated into the map, and using geographical location technology is able to provide the locus of an incident requiring action from the relevant agencies. The 'map' above is an actual digital map of the Kibera slums of Nairobi.

Figure 12.2 shows live post on the *Voice of Mathare* platform under the section 'reports'. This was taken on the actual day of voting, on April 3, 2013. Ideally, it was meant to use local voices and residents of the Mathare slums to capture incidents of violence or other disruptions to the poll on the material day of voting.

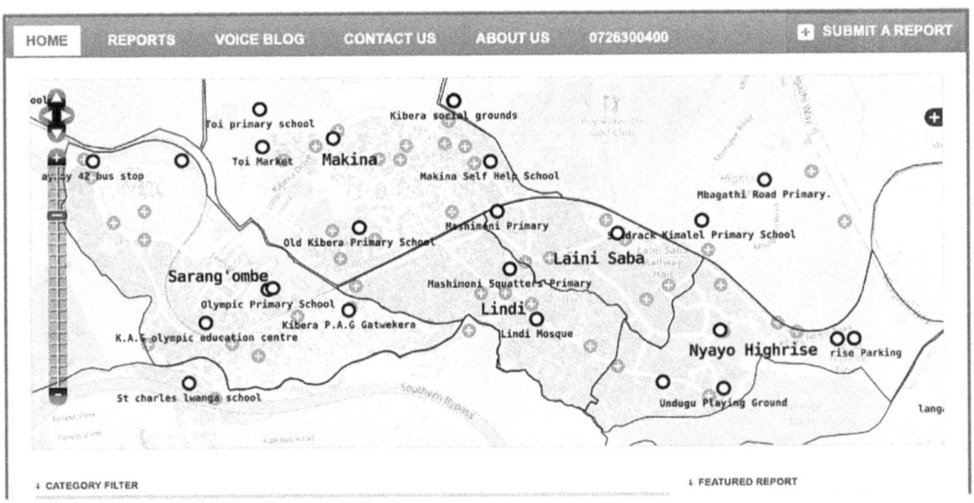

Figure 12.1 Voice of Kibera Crowdsourcing Platform.

commotion between ODM and FORD-K oficials ODM and FORD k agents at huruma polling station had a commotion as the ODM agents tried to mislead a FORD K voter.FORD K then claimed that ODM wanted...	2013-03-04	huruma primary school	YES
IEBC laptops having mechanical problem Voters are stranded at Utalii ward drive inn polling station the machine has some problem whereby the machine is not displaying names of the voter	2013-03-04	Drive inn	YES
BLAME TO IEBC OFFICIALS IN MATHARE SOCIAL HALL Voters are blamming the IEBC officials as voting at the mathare north social hall is at a slow pace.This is raised after voters queuing like for more...	2013-03-04	mathare north social hall	YES
voter turn out at St. Theresas girls primary and secondary school High voters have turned out at St Theresa girls primary and secondary as voters are still trickling in large numbers.	2013-03-04	St. Theresas	YES
robbery incidence a woman has been robbed in broad daylight at kiboro primary school to st theresa girls secondary and primary school by a group of young men masquerading...	2013-03-04	kiboro primary school	YES
kiamaiko ward voters come in large number voters at valley bridge primary and nduruno primary polling stations in kiamaiko ward turned up in large numbers to vote 4dea preffered candidates.	2013-03-04	kiamaiko	YES
police dispersing youths who are disrupting peace police have up brought Sanity at St. Stephens polling center in ruaraka constituency after youths were making noise disrupting peace in the area.	2013-03-04	st stephen	YES
arrangement begins St. Stephens polling at ruaraka constituency voters are being arranged in streams according to alphabetical order,Voting has not yet started.	2013-03-04	st stephen	YES
pushover leading to injuries Two ladies got minor injuries at Daima polling station in huruma ward at 6a.m.voters pushed the gate because they wanted to vote early and went inside...	2013-03-04	daima primary school	YES

Figure 12.2 Voice of Mathare, 'Live posts'.

'Urban poverty,' the 'Invisible End User' and the myth of peacebuilding in the digital age

Kibera has often been a favourite journalistic and academic metaphor of 'describing urban Africa', and more precisely, the ultimate proof of 'knowing Nairobi'. Once described as the largest urban slum in Africa, the informal settlement in the Langata area of Nairobi has a population of well over 200,000 people. Alongside Mathare, another informal settlement, Kibera, has become a favourite of anthropologists, tourists and civil society groups in a rising phenomenon of 'slum tourism', 'new forms of gaze' and a rising economy of 'compassion' that has made Nairobi the unofficial capital of Africa's civil society and aid groups. To experience Kibera, for non-locals, is a kind of achievement, a conquest of sorts, and to be seen as helping and working for residents of Kibera or similar other informal settlements in Nairobi and nourishes social and cultural capital that can be swapped for several other material and non-material benefits.

During elections, rarely would a story on Kenya by a western journalist be written without the curious inclusion of an escapade to or from Kibera. Mathare, its look-alike on the opposite side of the city, serves similar metaphoric and material functions. However, many commentators allude to the increased spectre of violence in the slums as a reason for the increased attention from journalists, academics and 'techno-evangelists' as they sometimes prefer to call themselves. However, the slum in the Kenyan context also allows us to interrogate peacebuilding practice as a network of interests, ideologies, intentionalities and rationalities that are often elaboration outside of the purpose of 'peacebuilding'. In particular, we show how the 'invisibilisation' of end users in the deployment of digital technologies during and after elections in Kenya is correlated to the possibility of deploying the perspectives and models that motivate a good number of internal and external actors involved in building peace.

A report by Sambuli et al. (2013), undertaken to evaluate the efficacy of crowdsourcing in monitoring the 2013 elections, served as good entry point to our funded research. The report, published by insiders in the crowdsourcing and digital turn to peacebuilding, offered some insights into how the creators envisaged the efficacies of these platforms. In this report, it was observed that in the 2013 elections in Kenya, Twitter outdid active crowdsourcing platforms such as *Uchaguzi* on the amount of relevant and 'actionable data' although it had vast amounts of 'worthless' posts. The report made the assumption that information access through crowdsourcing is crucial to democratic growth in developing economies like Kenya. However, even with these claims, the report was largely silent on who were the actual end users of these platforms. As part of our research into how digital technologies played a role in peacebuilding in the 2013 elections, we were convinced that only through access to end users could we effectively evaluate the linkage between the intentions of administrators and funders of such platforms and their efficacies.

In a self-published report by the *Uchaguzi team* shortly after the elections, the absence of end users in a report whose objective was to measure impact was conspicuous. According to the reports, citizen reports were sent via SMS, twitter and Facebook, and as part of their recommendations, they argued for a countrywide study specifically focused on the users' perspective so as to understand how citizens use ICTs during elections and the opportunities and challenges thereof (Omenya and Crandall 2013, p. 30). Still, staff at I-hub research and the owners of *Uchaguzi* did share with us what they termed as 'raw form aggregations of citizen responses', which comprised of data collected from end users in three Nairobi slums and whose abridged form comprised part of the wider report. In the report, they argued that the three city slums were chosen because of the amount of information that was received from the locations and the historical context of the locations as perpetual hot spots for violence.

Respondents were selected through snowball sampling technique, and a total of 35 citizens were interviewed of which 26 were female and 7 were male. While all these reports make a contribution to understanding how Kenyans citizens appropriated crowdsourcing platforms in the 2013 elections, they failed to actually prove that those who were interviewed were the actual people who used these platforms. In actual fact, while these reports were apparently part of 'end user surveys', there was virtually no actual mention of or profiling or description of actual end users. Where were the end user of the *Uchaguzi* platforms and who were they?

According to several interviews with administrators and the research teams working with these platforms, legal issues to do with privacy were cited as a barrier to accessing the end users. They argued that an informal contractual agreement existed which stated that the identity and contact details of those who posted on the citizen reports cannot be shared out to third parties, like us. The details of these contracts were also not shared with us. This meant that we could neither identify and interview nor reach the so-called end users. We were persuaded that only actual end users could meaningfully provide us with details of what incidents they saw on the material day, how they interacted with the various interfaces and whether or not what was reported was responded to by the authorities. Further, we felt that linking government or emergency response to an incidence of violence directly to a 'citizen's report' online could only be verified by end users.

The three administrators of the *Voice of Kibera, Voice of Mathare* and *Uchaguzi* platforms informed us that a plug-in instantly anonymised posts coming in from the field, meaning that they themselves could not know who actually posted in these platforms. In this sense, while substantive information could be relayed and shared with concerned authorities, the identity and personal details of the particular citizens remains anonymous. As such, we were unable to really establish who really these end users were and their motivations and the manner of interface with these platforms. Still, it was very clear that since many of the published reports were by persons directly or indirectly affiliated with the leading crowdsourcing platforms, there was a subtle yet recognisable attempt at glossing through critical attributes of the end users, and at the same time a clear imperative at extolling the space and performance of crowdsourcing platforms.

In further interviews seeking information on use, we wanted to find out from the administrators what their actual end user categories were, and how they went about settling on particular target groups and not others. We found that the roles of what one would call administrators seemed to be a bit ambiguous, with some administrators—for reasons unknown to us—shifting responsibility of administration to entities that are impersonal (i.e. groups such as youth organisations that are domiciled in slum communities). For the *Voice of Kibera*, the administrators indicated that theirs was an improvement of an initial mapping of Kibera, only that this time there was the intention to 'add narrative to the spatial map information in real time'.[7]

As for the *Voice of Mathare*, the administrator developed it from a previously existing blog, and its aim was to capture newsworthy events in the slums. Not surprising, our inquiries on the actual profiles and motivations of end users elicited blank responses. It was apparent that all the three platforms had links to external western funding, with most administrators being westerners. We observed that digital technologies in Kenya obtained a peculiar political economy of technical assistance and external funding in the context of peacebuilding. There is, on one hand, an external demand for 'stories' that stand out in compliance with expectations that 'resilience', 'innovation' and 'resistance' take specific forms in African settings. On the other hand, civil society organisations and the perceptive public have become keen to capture the market for external attention and they have perfected strategies of response to their own understanding of what is expected of them.

After further fieldwork within Kibera itself and discussions with several youth groups associated with digital platform used during the 2013 elections, we concluded that the so-called end users were not an anonymous mass, indeterminate public unknown to each other. Rather, most 'end users' were part of a larger insider network that linked the plethora of externally funded 'youth groups' in slums with administrators of crowdsourcing platforms, where for a 'modest incentive' the 'youth' would populate the platforms with information on what was happening in several sections of the City's slums during the election period. We concluded that the deployment of crowdsourcing platforms in the city's slums during the 2013 general elections had some disturbing patterns of artificiality.

Regarding *Uchaguzi*, which was ostensibly deployed as a countrywide 'elections mapping platform', we correctly assumed the same since the 'citizen reports' in *Uchaguzi* emerged from across the country; we did a survey involving members of the public and cyber café owners in the centre of Eldoret town (a town west of Kenya with a population of about a million people). Eldoret had witnessed the most extreme form of violence in the 2007 general elections and the *Uchaguzi* platform had numerous citizens' reports coming from Eldoret.

Our intention was to gauge awareness levels of *Uchaguzi* platforms in Eldoret. We narrowed the study to owners of the seven of the busiest and most frequented cyber cafés in town. We wanted to find out from these operators whether they knew anything called *Uchaguzi*, and whether any of their clients accessed these platforms. Of the seven cybercafé owners and users, none had heard of *Uchaguzi* nor seen any of their clients talking or seeking help to access the platform. We were not entirely surprised with these findings, since these platforms were domiciled, publicised and appropriated mainly within the city of Nairobi. Still, we were unable to account for the many reports coming from the city of Eldoret that were published in the platform.

Again, citing legal and other reasons, there was very little that administrators could do in giving us contacts or leads of the actual end users. What we found, however, from interviews with administrators was that what they termed 'strategic partners', 'NGO colleagues' and other closely linked entities formed a significant portion of these end users than what would be the ideal, common man end user. In actual fact, creators of crowdsourcing, mapping platforms designed to monitor, secure and build peace during Kenya's elections week were largely comprised of a network of people belonging to a similar civil society/NGO field, where participants, known to each other, understood the logic required in making the platforms work, or be seen to be working, and where the assumed success was understood and articulated as a generalised success to all players, most of whom are 'based' in Nairobi's slums through organised groups. This category of 'end users' was unlikely to be reliable in investigating the efficacies of digital media in peacebuilding.

While some of the administrators argued that this category of 'induced end users' were essential in getting the platform take off and crucial in ensuring its very sustenance, it was an obvious contradiction of the articulated claims of helping marginalised groups and borderline social categories participate in civic affairs. As a result of these dynamics, the researchers felt constrained to innovate a category of 'simulated users'. But we felt that such responses are not only unethical, but is more likely to be predictable, 'sexed up' and several times removed from actuality. Interestingly, in one of their own published reports, the evaluators of the *Uchaguzi* platform appeared to question the whole rationale of seeking out actual end users. In a published report, the creators of the crowdsourcing platform argue that it is far better to work with online discourse than perform actual fieldwork studies in seeking to establish and account for the profiles of end users. They argue that it is easier to find eyewitnesses (read end users) online than in person. In a follow up in Kisumu, the 'in-house' researchers claim they encountered barriers similar to the ones we faced:

> During the fieldwork we conducted in Kisumu, our research team actually found it very difficult to obtain first-hand or even second-hand accounts of violence. Although many of our 85 interviewees acknowledged witnessing some forms of violence, often their accounts lacked the detail and even the personal quality of posts (online)… As a result, we concluded that verifying incidences by in-person interviews with victims

> and residents of areas where violence had been reported (online) was not necessarily the most effective method. It was also difficult to find the appropriate individual who might have witnessed an event.
>
> (Omenya and Crandall, 2013, p. 40)

What then does this mean to researching the digital with regard to peacebuilding in Africa? Without credible data of how ordinary users engage with these innovations, it is impossible to make any claims with regard to the successes or otherwise of the 'digital turn' in peacebuilding. From this experience we gather that the pervasive language of peacebuilding might blind us to its disciplining effects. What we are being made to research may not always have much to do with peacebuilding but interventions that may or may not provide useful insights into peacebuilding experiments. Further, focused primarily on elections, digital technologies anchor a liberal peacebuilding approach that assume that perceptibly 'good elections' and open market systems produce 'good things' (Curtis and Dzinesa, 2012). In this case, the ghosts of the 2007 post-elections violence, on one hand, and the euphoria around *Ushahidi* and its many spin-offs, on the other hand, nourished a space where discourses of digital technologies for monitoring elections flourished, even when the relationship was tenuous.

Further, this research suggests that critical questions have to be raised in the 'digitalisation' of 'peacebuilding', and there is need to flesh out the rising motivations around technology digital/ICT export and digitalisation programme. In particular, it is important to assess the colonising/neo-colonial effects of the current flow of technological innovation to peacebuilding in Africa. In the case we have highlighted, it is important to go beyond motivations and probe why communities assumed to be marginalised and excluded from digital innovations—slum communities—are further excluded from programmes that are assumed to be directly drawing from and speaking on their behalf.

Conclusion

The textual analysis of much of the posts displayed in the three platforms under study, demonstrates that while the target user was the average slum dweller in Nairobi, many of the reports did not actually come from such a category. The absolute impossibility at reaching or contacting actual end users only complicated the situation. While it is possible stringent privacy requirements from donors to the creators of these platforms prevented access to end users, the acknowledgement from administrators that a large number of 'insider' postings were 'induced' raises questions of the efficacy of digital media to peacebuilding initiatives. Also, in deprived, poverty stricken urban communities, the use of 'small' amounts of cash to drive up activity in such platforms might lead to a performance that is more deliberate than natural.

However, this does not mean that the digital cannot be effective in monitoring elections-related violence or of digital media playing a role in peacebuilding in Africa. In a previous study done in Kenya (Omanga, 2015), the use of twitter was shown to be quite effective in community policing in sections of Nakuru, Central Rift Valley, Kenya. The main difference in the Nakuru case was that the innovation deploying twitter as a tool, used by local chiefs and connected to the residents of the respective administrative units, was initiated from below. It had local origins, and was developed with the local actors and with local realities in mind. In the case of the crowdsourcing platforms in the 2013 elections, the heavy foreign funding, alongside an abiding 'silicon savannah', 'Africa rising' discourse exalting the promise of digital technologies in Kenya—NGOs, youth groups and civil society groups—operating in and within slum communities took the cure and produced a 'digital peace' that served more the piper's payers vision of peace.

Notes

1 https://en.wikipedia.org/wiki/Ushahidi (Accessed November 11, 2018).
2 Africa rising. *The Economist*, December 3, 2011.
3 See http://ca.go.ke/images/downloads/STATISTICS/Sector%20Statistics%20Report%20Q2%20%202017-18.pdf
4 http://kw.awcfs.org/article/kenyans-among-top-100-most-powerful-women/
5 https://ihub.co.ke/blogs/12080/uchaguzi-meet-up-electionske-2013-and-uchaguzi
6 https://ihub.co.ke/blogs/12080/uchaguzi-meet-up-electionske-2013-and-uchaguzi
7 Source: Interviews with *Voice of Kibera/Voice of Mathare* 'administrator', December 2013 (Nairobi).

References

Asiedu, C. (2012). 'Information Communication Technologies for Gender and Development in Africa: Thecae for Radio and Technological Blending.' *International Communication Gazette*, 74(3): 240–257.

Berger, G. (2002). 'Theorizing the Media-Democracy Relationship in Southern Africa.' *International Communication Gazette*, 64(1): 21–45.

Curtis, D., and Dzinesa, Gwinyayi, A. (2012). *Peacebuilding, Power, and Politics in Africa*. Athens: Ohio University Press.

Molony, T. (2018). 'Social Media Warfare and Kenya's Conflict with Al Shabaab in Somalia: A Right to Know.' *African Affairs*, 118(471): 1–24.

Mudhai, F., W. Tettey, and F. Banda, eds. (2009). *African Media and the Digital Public Sphere*. New York: Palgrave Macmillan.

Ndemo, Elijah Bitange. (2015). 'Political Entrepreneurialism: Reflections of a Civil Servant on the Role of Political Institutions in Technology Innovation and Diffusion in Kenya.' *Stability: International Journal of Security and Development*, 4 (1): pp. 1–14.

Nyabuga, G. (2008). *Click on Democracy: Uses and Effects of the Internet on Kenyan Politics*. Saarbrucken: VDM Müller.

Omanga, D. (2015). '"Chieftaincy" in the Social Media Space: Community Policing in a Twitter Convened Baraza.' *Stability: International Journal of Security and Development*, 4 (1): pp. 1–16.
Omanga, D., and Pamela Chepngetich-Omanga. (2013). 'Twitter and Africa's "War on Terror": News Framing and Convergence in Kenya's Operation Linda Nchi.' In A. Olorunnisola and A. Douai (eds.), *New Media Influence on Social and Political Change in Africa* Hershey, PA: IGI Global, pp. 241–256.
Omenya, Rhonda, and Crandall, Angela. (2013). Uchaguzi Monitoring and Evaluation Report, p. 30. http://whiteafrican.com/wp-content/uploads/2013/03/uchaguzi-kenya2013.pdf (Accessed November 11, 2018).Postman, Neil. (1970). The Reformed English Curriculum. In A. C. Eurich (ed.), *High School 1980: The Shape of the Future in American Secondary Education,* New York: Pitman, pp. 160–168.
Price, M., and N. Stremlau. (2012). 'Media and Transitional Justice: Toward a Systematic Approach.' *International Journal of Communication*, 6 (2012), pp. 1077–1099
Scolari, C. A. (2012). 'Media Ecology: Exploring the Metaphor to Expand the Theory.' *Communication Theory*, 22: 204–225. doi:10.1111/j.1468-2885.2012.01404.x.
Tacchi, J., D. Slater, and G. Hearn. (2003). *Ethnographic Action Research, UNESCO Users Handbook*. Retrieved from http://eprints.qut.edu.au/4399/1/4399.pdf.
Ugangu, Wilson. (2011). 'The Silences in Public Debate Created by Lack of Freedom in Kenya.' *African Communication Research*, 4 (2), pp. 247–267.

13 Post-conflict emotions at home

Researching former Zimbabwean soldiers

Godfrey Maringira

Introduction

This chapter focuses on the ways in which I engaged with the former soldiers as research participants in the aftermath of political violence in Zimbabwe. Researching soldiers who have lived their lives as both victims and perpetrators of political violence is challenging because their stories are characterised with mixed emotions of that violence. What is more unsettling is the ways in which research is done by an insider: one who has been working with this category of people. I argue that when researchers study the people whom they had lived with, the research in itself reproduces 'home' and/or 'familiar' identity. In other words, the ways in which researchers doing research at home access the field of study, the questions they ask, the way they 'hang-out', the kind of expectations they have from informants, and the emotions carried on throughout the field produce that which is self-ingrained in the past identity.

This research made me feel like I was reliving the moments of the past violence again. I was fascinated and disturbed by my participants' stories of their experiences of political violence and being in particular places I knew. In this chapter I reveal the struggles to shrug off 'home identity' and emotional attachment of the people and places we know when doing the research in post-conflict. Thus I contend that when doing research, talking with your community and being among the people you know is both a resource and a challenge.

Soldiers and violence in Zimbabwe

The history of political violence is the history of soldiers in Zimbabwe. Since independence, in 1980, Zimbabwean soldiers have been involved in political violence against their own citizens, and have perceived them as political opponents (Alexander, 1998; White, 2007; Sachikonye, 2011; Alao, 2012). Soldiers have been and continue to be the hallmark of politics and violence. The former president, Robert Mugabe, has made it clear that the military anchors the state when he reiterated the function of senior army officers in previous elections. This therefore positioned soldiers in politics, which scholars have noted that led to the 'militarisation of politics' (Alexander, 2013), the 'politicisation of the military' (Masunungure, 2011; Rupiya, 2011) and the 'militarisation of the military' (Maringira, 2016; Maringira, 2019). Such analysis of the military helps us understand how the military involvement in politics defines their work of political violence.

Soldiers have been heavily involved in the historic civilian massacres immediately after the Zimbabwean independence (CCJP, 1999). In addition, they have also been perpetrators of political violence in the post-2000 during the land reform programme which saw the eviction of white commercial farmers from their farms (Chaumba et al., 2003). Again, in all preceding elections, soldiers' use of violence has been evident, but with no sustainable solution to the problem engulfing the country. What has also been evident is that soldiers have been victims of the same violence, which the state meted on civilians (Maringira, 2017a). However, this chapter does not explore these issues as they have been explored elsewhere (Maringira et al., 2015; Maringira 2017b). Instead, it details how I engaged with these former soldiers in the aftermath of doing violence and experiencing it as victims in the same process.

Entering the post-conflict field

On re-entering the field site, the community, the former soldiers and I intuitively greeted each other. I was familiar with the soldiers when they were perpetrators of political violence, and their people, houses and families. We talked about our past lives, how the present was affected by the past and the ways in which the present can shape the future. We literally and figuratively turned the fieldwork into a home space, in which both the past and present was shared as well as our aspirations for the future. I ended up thinking about and doing this research among the former soldiers I had known. My previous relationship with and knowledge of these soldiers made the research feasible. I understood the rationale behind the events narrated by the interviewees. The former soldiers largely perceived me as a sympathetic listener in whom they could confide. This allowed them to speak without censoring themselves and to divulge their experiences of violence as well as their experiences in the aftermath of violence.

Despite this, I had some challenges in making my participants understand what I was researching and why I was interested in their past and current lives in the aftermath of political violence. This dilemma was a result of the fact that in Zimbabwe people were quite paranoid and anyone asking about political violence and the ways in which people lived through it is considered a 'spy'. Most of the participants were either perpetrators or victims of violence. The participants were concerned about the possibility that the information I was gathering would be made public and fall into the wrong hands. They feared that the Central Intelligence Organisation (CIO), and the Zimbabwe Military Intelligence – that fearsome Zimbabwe government spy agency – could use knowledge about them against them.

I had to be aware all the time that my participants were fearful of the Zimbabwean state spy agents. To talk about my research project, I used the term *tsvakurudzo*. In Chishona, it literally means one is seeking to investigate and understand people's lives. In Zimbabwe, this is mostly done by intelligence officers. By saying that I wanted to do *tsvakurudzo*, I had momentarily transformed myself into an intelligence officer meant to get them. However, through extended explanation of what academic research means, and with reference to my university as well as by showing them my credentials (letter of registration, ethical clearance letter and identity card from the university), they understood the parameters of academic research and its difference from 'spy work'. Of course, they remained unconvinced, though they consented to participate in the research.

After ensuring their informed consent, from 2011 to 2012, I participated in my participants' activities for purposes of immersion in their social world and as part of building my fieldwork relationship with them. I communicated with the research participants by cell phone, Facebook chats and WhatsApp. I accompanied some of them to their work places which were mainly in the informal labour market. I visited them in the streets where they worked as street vendors. At one point, I had to help one of the participants to sell his wares, and I took the opportunity to have conversations and pose some questions which would not be possible if I had remained idle or a visitor to his 'street office'. I observed that the Municipal Police were harassing vendors selling in the street. As the Municipal Police approached, my participant, Tongai, introduced me as one of them. I later asked him why he was 'telling the municipal police that I am one of them, when in fact I was not as participating in the community as before'. Looking surprised, he said: 'you are still one of us! Even here where I am selling airtime, some municipal policemen know me very well!' I realised that Tongai was using his community belonging and perceived entitlement rights to claim space in the city.

Despite these statements by participants affirming my membership of their community, I maintained the 'invented', 'temporary' and 'artificial' distance and position of researcher, sometimes with great distress. Oscillating between the spaces and experiences of being at home with them as an ordinary member of the community and a researcher allowed me to deal with, what Bourdieu (1990, p. 124) calls two incompatible points of view, objectivism and subjectivism: how to understand the subject within objective structures. It is therefore difficult for one to claim objectivity when in actuality it is clear that subjectivity is evident in the text. One of the moments when the clash between objectivity and subjectivity became evident was when one of the participants, Tafara, sent me a cell phone text message, saying:

> We want to issue a statement condemning military involvement in politics, may you urgently contribute?
>
> (Tafara, 17 August 2012, Johannesburg)

The message revealed how important I was to the former soldiers. I did not respond, but kept the text message on my phone. It reminded me of the ethical issues related to my study and the community awareness that the soldiers had no support to lobby for the alleviation of their plights. It also showed how they perceived me as tied to their agenda. To a degree, in light of my experiences and as a researcher, which I shared with them, I was emotionally involved with this group.

Even though I wanted to maintain my scholarly and research position, they continued to look up to me for counselling interventions, especially those who were experiencing emotional distress and violence-related nightmares. I referred those who needed counselling to organizations that provided free services. Most of them refused. They regard these institutions, staffed by professional psychologists and counsellors but with no home and/or community background, as ill-suited to deal with their issues. Nonetheless, I did not counsel any of the participants in my study. I felt that such an intervention would be contrary to the ethical undertaking I had given to the university that in case where my participants need counselling, I would refer them to professional counsellors.

Reproducing the past through story-telling

Initially, a structured questionnaire was used to solicit responses to people who were affected by violence. It soon became clear that whenever a question was asked about post-violence lives, most would dwell on their life experiences, going back to the violent moments. They delved into the events that led to political violence. This showed that the use of a life story approach was more suitable, especially coupled with follow-up in-depth interviews. The life stories enabled me to trace how violence began, the violence they experienced and how they are living in the aftermath of it.

In retelling their stories, they revealed their experiences at four levels: firstly, *as lived* – the events that touch on a person's life. Second, the related life *as experienced*: how a person perceives and ascribes meaning to what happens. The third level involved their lives *as told*: how experience is framed in a given context to a particular researcher. Finally, life as *a text*: interpreted and presented by the researcher (Denzin, 1989; Eastmond, 2007). My personal experiences and home assumptions acted as filters in which the stories were subsequently woven and produced into a meaningful text.

Life histories produce what Bourdieu (1990, p. 124) calls an 'account of the accounts', i.e. it produces detailed stories, in this case that of political violence and its aftermath. Many of the life history interviews lasted for three hours, with the longest lasting for four hours and producing a 50-page transcription text. At times, especially when the participants became distressed, we had to break and resume the interviews on other days. Sometimes the participants simply had too many experiences to discuss. Then we would resume, e.g. after four hours of talking, on another day. Life histories evoke stories and memories that would not be possible to share in ordinary conversations (Dhunpath, 2000; Bruner, 2004; Denzin, 1997).

The participants told the stories differently. They narrated it in ways they believed were meaningful and revealing of their experiences. Afterwards some felt good while others were disturbed. In such cases they often asked to continue the discussion on social media at night. Our social bonds were reconfigured within these interactions. I was able to follow my participants' stories because I understood the context of events being narrated. I found the research productive to me as a learning process, which gave me an opportunity to find some perspectives of the past of which I am a product and also a part. The stories of violence and its aftermath reveal that the participants' experiences of violence constitute the greater part of their personal biographies. I argue that when doing peacebuilding research, people do not always talk about peace, but their narratives are those of violence. Researchers usually think that when doing peacebuilding research the narratives that would emerge will be those related to peace. In fact, much of peacebuilding research constitute of narratives of violence.

For Eastmond (2007), stories in life histories reconstitute identity. While Eastmond (ibid.) was referring to the identity of the participants, in my study, my own home identity was reproduced both consciously and unconsciously as I negotiated my way into and through my participants' lives. The struggles that I encountered and still encounter as I move between the social worlds of being at 'home' and that of a researcher sometimes leave me stuck betwixt and between in contradictory positions. For Goodson (1992), the story of his life is always embedded in the story of those communities from which he derives his identity.

Reflecting on reflexivity

It will be misleading to state that I have managed to escape from my own home identity to understand my current research. Instead I used my home identity and the familiarity of the place to discuss the stories told by the participants. I have historical and emotional ties to the community I studied. My past continues to haunt me, and this is represented in the ways in which the text has been interpreted. According to Tankink and Vysma (2006) the subjectivity of the anthropologist influences the research process. This is particularly the case for a researcher who shares similar experiences with their research participants. Below, I discuss issues related to reflexivity and how I situate my own involvement in this study. I will start by reflecting on reflexivity in a broad sense with regards to research, and then I will narrow it down to 'reflexivity and researching the familiar group' (see also Higate and Cameron, 2006).

In doing this kind of study I reflected on my own home life. This made me understand that being reflexive is not only about my past experiences at home, but also about the way I accessed my participants, and the relationships I forged and maintained in and outside the field with my participants. Borrowing from Giddens' (1991) idea on the 'temporal dimensions of reflexivity', I have also realised that reflexivity is not temporal but moves beyond the fieldwork phase. Similarly, Riach (2009) emphasizes that reflexivity is embedded *in* and *after* the moment. This is what Finlay (2002) refers to as 'thinking about'. In regard to this, Hall and Caller (2001) argue that when collecting data, the researcher will be doing it with participants and he/she will also be in conversation with the interview scripts when writing.

Scholars on reflexivity question the epistemological and ontological issues related to understanding the social world and diverse views. They argue that the social world is not an object to be discovered; rather, researchers actively construct reality as they study it on the basis of their epistemological and ontological assumptions (Cunliffe, 2003). While I was knowledgeable of the stories these former soldiers tell, I also learned of certain issues which I initially took for granted. How we think shapes the research, how we formulate our research questions the data collection tools to use and the stories to include in the data analysis. Reflexivity recognises the influence of the researcher's values and assumptions on the process of inquiry (Finlay, 2002; Cunliffe, 2003). Thus *being* reflexive 'is about the interpretation of data through which meanings are drawn rather than found' (Mauthner and Doucet, 2003, p. 414).

This means doing research on men who had been involved in violence at home had its epistemological and ontological challenges. In order to explore this, I lean on Bourdieu and Wacquant (1992) who argue that this kind of reflexivity is drawn from 'practical mastery', which is derived from one's past experiences learned and inculcated in one's individual being and practice. There is also 'symbolic mastery' (Bourdieu and Wacquant) in which the researcher draws on their subjective experience. Nonetheless, Alvesson et al. (2008) argue that it is not only about carrying the past self in the field but recreating it in conversation with participants. The home culture creates insider and outsider categories vis-à-vis researcher and the interviewees (see also Higate and Cameron, 2006). Reflexivity enhances methodological rigour and it is at the centre of creating opportunities for researchers, researching at home as it enhances the 'deep' hanging-out with the participants (ibid.). For Bourdieu and Wacquant (1992), it is the 'social condition' that forges a relationship between the researcher and the participants. For Schmidt (2007), what we know informs how we act in specific contexts. This is what Pierre Bourdieu (1990) calls 'epistemic reflexivity', which reveals our subjective experience forms a particular kind of analysis. Reflexivity is therefore a methodological and analytical resource (Higate and Cameron, 2006).

While I endeavoured to minimise bias, it will be an epistemological fallacy to say that I have detached myself from the data gathering process and analysis. The motivation to engage in this study was premised on the past experiences and the ongoing relationship between my research participants and myself. While Hardy et al. (2001, p. 536) talk about how subjects are produced in the research process, when my participants were telling their stories in an interview it was not only a time of remembering what they did and how they did it, but it was also a time for me to reflect on the community. I also shared with them a kind of home-ness, i.e. a sense of and feeling at and being at home. It viscerally reminded me of violence in the community. It brought me back to how we survived the violence as well as experiences of being victimised.

While reading the transcripts, I felt like the research had 'deployed' me back into the violence community. Though these thoughts constantly entered my mind, it was also fascinating to re-read both transcribed texts and the previous experiences in which the participants depict themselves as 'community heroes' surviving political violence. It is a past that I now – retranslate and rewrite through the participant experiences produced and filtered through an anthropological study in an academic space.

A researcher's biography plays a significant role in the whole research process. This is well supported by Roberts and Sanders (2005, p. 294), who demonstrated 'how biography and [community] knowledge are two resources that influence research practice'. The research process allowed me to understand the life of the community. I came to realise that I am constantly torn between two contexts in a bid to understand myself. So, the self, including myself, is always negotiated. The ways in which I probed questions during interviews reveal a degree of prior knowledge of the events narrated. The same applies to what I wrote in my fieldwork diary. Bourdieu (1990, p. 33) classifies this as subjectivity research, which reveals how prior knowledge is what steers us to delimiting our research work.

There were also ongoing power tussles between my participants and me when I was interviewing and interacting with him. This is what Bourdieu (1990, p. 127) calls a 'field of power', which he explains as 'positions in the space of power position'. I experienced 'momentary reversion' to the role of an outsider (Higate and Cameron) and occupied a subordinate position. For Pierre Bourdieu (1990, p. 127), 'the interactions, which are accepted at their face value by people...one can observe them, film them, record them... conceals the invisible which determines it'. I was constantly reminded of my past life in the community.

Interviewing the former soldiers on the violence which they had experienced momentarily reintroduced our past relationship – that of platoon commander and section commander. The 'social facts' (Durkheim, 1973), which defined our community relations for a moment, constrained my behaviour. There was a persistence of debasement of my past, which continued when I was doing fieldwork. For Bourdieu (1990, p. 127), this is akin to 'strategies of condescension' by which agents who occupied a higher position maintain social distance. In this regard, the research reminded me of my social position.

Thinking through the research

Throughout the writing process, the analysis drew me closer to the community experiences of political violence. The process of data analysis proved to be a tedious process, which began with transcription and re-listening of the transcribed data to do a thematic analysis (Braun and Clarke, 2006). At the end of the writing process, I felt I had become more re-embedded into the community life than before, as I went back and forth through the data I had collected and made sense of it.

The approach that I adopted to read the transcripts was to try and have a closer reading of the interviews alongside existing literature. When reading I could always relate to my 'community past'. The reading of transcripts helped me develop an in-depth understanding of transcripts, especially to draw themes emanating from the participants' voices and to rethink and use the often taken-for-granted terms that I used in my writing. While the kind of themes, which I presented in writing elsewhere in other papers, was motivated by participants' accounts, my own 'community reflection' played a part. All this sometimes happens unconsciously and one only notices it in rereading the text. It is a practice, which at the beginning of doing research at home one would think of avoiding but somehow it becomes unavoidable. I also followed what Fereday and Muir-Cochrane (2006) called the 'thematisation of meanings', which implies that the meanings derived from the data is categorised into themes.

The writing process was not very different from a continuous process of reading and rereading of transcripts in the sense that I continuously reflected on the data collection process. The study itself, as well as the process of listening to the tapes, transcribing them, rereading, thinking about and analysing my data was very evocative and even disturbing. The things my participants told me haunted me. Sometimes, it would come back in my dreams as I reflected on my fieldwork and writing. To avoid nightmares, I began to work during the night and slept during the day, and this recurred throughout my writing process. It was a choice for me that worked.

Conclusion

Doing research in post-conflict contexts, especially among soldiers and at home, has its methodological benefits and setbacks. The data collection and writing process made me reflect more than before on what the political violence did to people I knew and the places I was familiar with. The violence narratives produce particular social and emotional attachment to our own homes and places we do research. At first, I was more concerned with just going to the field and collecting data, not knowing that I would be immersed in my thoughts of home, and in shared memory and history of the community. Who committed this act? And why? What are the consequences? The researcher and participants grapple with these questions, with one trying to get answers and the other trying to dodge them. No amount of awareness or preparedness can strip the knowledge the researcher has about his or her community and its people.

References

Alao, A. (2012) *Mugabe and the Politics of Security in Zimbabwe*. Montreal, QC: McGill-Queens University Press.

Alexander, J. (1998) "Dissident Perspectives on Zimbabwe's Post-independence War, Africa," *Journal of the International Africa Institute*, 68(2), pp. 151–182.

Alexander, J. (2013) "Militarisation and State Institutions: 'professionals' and 'soldiers' Inside the Zimbabwe Prison Service," *Journal of Southern African Studies*, 39(4), pp. 807–828.

Alvesson, M., Hardy C. and Harley B. (2008) "Reflecting on Reflexivity: Reflexive Textual Practices in Organisation and Management Theory," *Journal of Management Studies*, 45(3), pp. 480–501.

Bourdieu, P. (1990) *In Other Words: Essays towards a Reflexive Sociology*. Stanford, CA: Stanford University Press.

Bourdieu, P. and Wacquant, L. (1992) *An Invitation to Reflexive Sociology*. Cambridge: Polity Press.

Braun, V. and Clarke, V. (2006) "Using Thematic Analysis in Psychology," *Qualitative Research in Psychology*, 3(2), pp. 77–101.

Bruner, J. (2004) "Life as Narrative," *Social Research*, 71(3), pp. 691–710.

Catholic Commission for Justice and Peace and Legal Resources Foundation, (1997) Breaking the Silence: Building True Peace: Report on the Disturbances in Matebeleland. Harare: CCJP and LRF.

Chaumba, J., Scoones, I. and Wolmer, W. (2003) "From Jambanja to Planning: The Reassertion of Technocracy in Land Reform in South-Eastern Zimbabwe?," *The Journal of Modern African Studies*, 41(4), pp. 533–554.

Cunliffe, A. L. (2003) "Reflexive Inquiry in Organisational Research: Questions and Possibilities," *Human Relations*, 56(8), pp. 983–1003.

Denzin, N. K. (1989) *Interpretive Biography*. Qualitative Research Methods Series 17. Newbury Park. London: Sage Publications.

Denzin, N. K. (1997) *Interpretive Ethnography: Ethnographic Practices for the 20st Century*. Thousand Oaks: Sage Publications.

Dhunpath, R. (2000) "Life History Methodology: 'Narradigm' Regained," *Qualitative Studies in Education*, 13(5), pp. 543–551.

Durkheim, E. (1973) *On Morality and Society*. Chicago, IL: The University of Chicago Press.

Eastmond, M. (2007) "Stories as Lived Experience: Narratives in Forced Migration Research," *Journal of Refugee Studies*, 20(2), pp. 248–264.

Fereday, J. and Muir-Cochrane, E. (2006) "Demonstrating Rigor Using Thematic Analysis: A Hybrid Approach of Inductive and Deductive Coding and Theme Development," *International Journal of Qualitative Methods*, 51, pp. 80–92.

Finlay, L. (2002) "Outing the Researcher: The Provenance, Process, and Practice of Reflexivity," *Qualitative Health Research*, 12(4), pp. 531–545.

Giddens, A. (1991) *Modernity and Self-Identity*. Cambridge: Polity Press.

Goodson, I. F. (1992) *Studying Teachers' Lives*. London: Routledge.

Hall, W. A. and Caller, P. (2001) "Enhancing the Rigor of Grounded Theory: Incorporating Reflexivity and Relationality," *Qualitative Health Research*, 11(2), pp. 257–272.
Hardy, C., Phillips, N. and Clegg, S. (2001) "Reflexivity in Organisation and Management Theory: A Study of the Production of the Research Subject," *Human Relations*, 54(5), pp. 531–560.
Higate, P. R. and Cameron, A. (2006) "Reflexivity and Researching The Military," *Armed Forces & Society*, 32(2), pp. 219–233.
Maringira, G., Gibson, D., and Richters, A. (2015) "It's in My Blood" The Military Habitus of Former Zimbabwean Soldiers in Exile in South Africa," *Armed Forces & Society*, 41(1), pp. 23–42.
Maringira, G. (2016) "When the Military Became Militarised: Accounts of Zimbabwean National Army Deserters in Exile In South Africa," *African Security Review*, 25(1), pp. 21–30.
Maringira, G. (2017a) "Politicization and Resistance in the Zimbabwean National Army," *African Affairs*, 116(462), pp. 18–38.
Maringira, G. (2017b) "Soldiers as Victims: Behind Military Barracks in the Post-Colonial African Army," *African Security Review*, 26(1), pp. 77–86.
Maringira, G. (2019) *Soldiers and the State in Zimbabwe*. London and New York: Routledge.
Masunungure, E. (2011) "Zimbabwe's Militarized, Electoral Authoritarianism," *Journal of International Affairs*, 65(1), pp. 47–64.
Mauthner, N. S. and Doucet, A. (2003) "Reflexive Accounts and Accounts of Reflexivity in Qualitative Data Analysis," *Sociology*, 37(3), pp. 413–431.
Riach, K. (2009) "Exploring Participant-centered Reflexivity in the Research Interview," *Sociology*, 43(2), pp. 356–370.
Roberts, J. M. and Sanders, T. (2005) "Before, During and After: Realism, Reflexivity and Ethnography," *The Sociological Review*, 53(2), pp. 294–313.
Rupiya, M. R. (2011) *The Military Factor in Zimbabwe's Political and Electoral Affairs*. Harare: Crisis in Zimbabwe Coalition.
Sachikonye, L. (2011) *When a State Turns on Its Citizens: Institutionalised Violence and Political Culture*. Auckland Park: Jacana Media Publishers.
Schmidt, A. (2007) "I Know What You're Doing, Reflexivity and Methods In Refugee Studies," *Refugee Survey Quarterly*, 26(3), pp. 82–99.
Tankink, M., and Vysma, M. (2006) "The Intersubjective as Analytic Tool in Medical Anthropology," *Medische Antropologie*, 18(1), pp. 249–265.
White, L. (2007) "'Whoever saw a country with four armies?' The Battle of Bulawayo Revisited," *Journal of Southern African Studies*, 33(3), pp. 619–631.

Index

Note: **Bold** page numbers refer to tables; *italic* page numbers refer to figures and page numbers followed by "n" denote endnotes.

www.ingramcontent.com/pod-product-compliance
Ingram Content Group UK Ltd.
Pitfield, Milton Keynes, MK11 3LW, UK
UKHW020147200926
475702UK00016B/2580
* 9 7 8 0 3 6 7 6 2 9 8 6 1 *